D0592434

Ambient Commons

Ambient Commons

Attention in the Age of Embodied Information

Malcolm McCullough

The MIT Press Cambridge, Massachusetts London, England

MIT Press books may be purchased at special quantity discounts for business or sales promotional use. For information, please email special_sales@mitpress.mit.edu or write to Special Sales Department, The MIT Press, 55 Hayward Street, Cambridge, MA 02142.

This book was set in Franklin Gothic and Garamond by the MIT Press. Printed and bound in the United States of America.

Library of Congress Cataloging-in-Publication Data

McCullough, Malcolm.
Ambient commons : attention in the age of embodied information / Malcolm McCullough.
pages cm
Includes bibliographical references and index.
ISBN 978-0-262-01880-7 (hardcover : alk. paper) 1. Architectural design—Philosophy. 2. Information commons. 3. Computer-aided design. 4. Human-computer interaction. I. Title.
NA2750M35 2013
720.1'08—dc23
2012030277

10 9 8 7 6 5 4 3 2 1

for Kit

Contents

Preface

To sit down with a book, gentle reader, seems less and less what life is like. To be alone with your thoughts was perhaps never the usual human condition. Whether in a traditional tribal society or amid today's tastes for perpetual messaging, some people don't share so much as generate their thoughts, even their deepest feelings, through constant social connectivity. Solitary reflection may have always been the exception, a chosen path for a few, or for a few chosen hours in a busy life; introspection remains a difficult art. Fortunately, to sit with a book is to enter a dialogue in any case: unlike feeding among so many short, separate messages at once, it is to take up a longer transaction between reader and writer. It is also enjoyable: books exist not only for greater depth or nuance, but also for the sake of language itself.

Here the transaction is this: like a highway resurfacing project, *Ambient Commons* asks your patience with a temporary inconvenience for the sake of permanent improvement. The project seems justifiable enough: whether you call it "overload" or "superabundance," the flood of mediated information defines this era. This book adds a tiny (albeit dense) drop to that flood. For, as the philosophers explain, often the solution to too much information is still more information, not only for tagging and filtering, but also for making sense.

So with hope for improvements in how you cope with superabundance, *Ambient Commons* invites you to rethink attention itself, especially with respect to your surroundings. You may sometime want to look past preoccupation with devices such as smartphones, and to notice more situated, often inescapable forms of information, but this isn't just about stopping to smell the roses. At the very least, this book may help you justify greater attention to your surroundings as something better than nostalgia. There is a practical new sensibility here. It belongs to a cognitive revolution. So the next time you mumble, "Be here now!" at someone texting while walking in your way, this book could help you mean more by that.

You might also understand that anyone, in any era, may have felt overload. That seems safe to assume. The world has always been overwhelming, all the more so whenever such basic needs as food or safety haven't been met. If you believe the old truism that you can keep only seven things in mind at once, then perhaps people began sensing overload as soon as there were eight.

Yet today surely something has changed. Much more in the sensory field comes from and refers to someplace else. Much more has been engineered deliberately for the workings of attention. The seductions of personalized media seem quite the opposite of a numbing, monotonous din. They don't lead to overload so much as to overconsumption. They aren't confined to specific sites like the workplace or the literary salon (where ladies and gentlemen of centuries past complained of overload—from the pressure to memorize too much poetry). Perhaps the biggest change is their ubiquity: as you may have noticed, the world is filling with ever more kinds of media, in ever more contexts and formats. Screens large and small appear just about everywhere; physical locations are increasingly tagged and digitally augmented; sensors, processors, and memory are not just carried about in pockets and bags, but also built into objects in everyday life. All these augmentations increasingly connect. This isn't the clanking industrial city that led sociologists to emphasize distraction, nor the media monoculture that led them to situationist critiques of spectacle, nor the all-seeing Orwellian state that many assume to be the inevitable outcome of unchecked mediation. Today's embodiments of information have become something far, far more chaotic, often quite culturally fertile, with ever subtler cognitive appeal. The twenty-first-century arts are the arts of interface. But interface is no longer just about sitting at a machine. To describe how this new technology, these new surroundings, and this new outlook for attention have begun to interrelate, this book adopts a single name, "the ambient."

When it comes to making sense of what just happened, a book still works best, at least across any time longer than a moment ago. (There is Twitter for that.) Although slow to produce, a book may still be the best way to trace an enduring path through the ideas of a recent decade or even a recent century. Here in the twilight of print, it helps to remember that. Although a book can't interconnect ideas nearly so readily as the web, it can penetrate some of them better. As a longer form in a consistent voice, a book may improve individual access to a field of ideas from many sources. I wrote this one to find out, and not from a position of expertise so much as one where I could do some digging. I wrote to help others find out for themselves, and for people in different disciplines to find out about one another. Please don't read this to learn something new about your own field. Read it instead as one writer's inquiry. The word *inquiry* should keep coming up. To write is to learn. To write well is to bring others along, even through superabundance. Know that, in the quickly rising flood of data, something just happened for attention to surroundings.

Acknowledgments

Let me begin by stating how much *Ambient Commons* reflects my good fortune to work at the University of Michigan, one of the world's foremost research universities. I especially value its combinations of expertise in tangible interface, cognition and environment, social history of technology, and networked urbanism. In all these areas, I am an amateur interloper. My university not only tolerates but often distinctly values my being so. In particular, esteemed colleagues Michael Cohen, Caroline Constant, Paul Edwards, Robert Fishman, Margaret Hedstrom, and John Marshall have generously shown me the way. Taubman College, where I teach architecture, has consistently provided encouragement for this, a project quite far from its core mission.

My work on this book would never have begun without the remarkable exposure provided by so many travels, especially in the boom years 2005–08, in many ways the dawn of embodied information. My thanks go to more individuals and organizations than I can name here. For one, I thank the interaction design organization IxDA for its events and many of its thought leaders for reading the earliest outlines of this book. With apologies for omissions, I thank Robert Fabricant, Sanjay Kanna, Michiel de Lange, Dave Malouf, Greg Petroff, Robert Reimann, Michelle Tepper, and John Thackara for volunteering to read and comment on foggy 20-page outlines in 2008–09.

Work on what would become *Ambient Commons* began in earnest when I took a long-delayed sabbatical from Taubman College in January 2010 to be a guest of Berkeley Center for New Media, an interdisciplinary center at the University of California. Director and kindred spirit Ken Goldberg kindly made this happen just as all purses and doors were snapping shut after the economic crash. Lifelong friends Edward Lee and Rhonda Righter came up with the idea for this visit in the first place, and did much to ease my time on the ground there as well. Destiny Kinal rented me her writer's hideout with shelves full of great books and a fine view of the Golden Gate, and off I went. When, by summer, I had a real manuscript under way, generous colleagues and former students took time to read and comment. Again with apologies for omissions, my thanks to Zack Denfeld, Steve Harrison, Danny Herwitz, Jessica Hullman, Erik Hofer, Diana Khadr, and Noah Liebman for their insights.

Along the way, I had a great many conversations face-to-face and by email with a wide range of researchers; many of these are cited in the book's endnotes. Let me thank a few of those who were especially helpful: cognitive scientist Marc Berman, overload scholar Finn Brunton, philosophy educator Brian Bruya, journal editor Ulrik Ekman, urban informatics pioneer Marcus Foth, urban historian David Henkin, journalist (and attention expert) Maggie Jackson, computer scientist and design educator

Scott Klemmer, interaction designer John Kolko, linguist and information historian Geoff Nunberg, sensate city researcher Dietmar Offenhuber, data formation master Casey Reas, ambient media entrepreneur David Rose, design research strategist Paul Siebert, attention teacher Lynette Smith, and software attention expert Linda Stone. My thanks as well to the Science-Technology-Society reading group in my university, and to the legendary design futurists at Steelcase for all their encouragements. I hope that none of you is too embarrassed by the result.

Thanks to the MIT Press for taking this unconventional project on. I must confess that, amid dealings with its own case of information superabundance, the Press knows how to let a text age, like wine, and that this proves wise in the end. Indeed it can be just what goes missing among others who hastily publish online. Across these cycles, somehow it was January, two more of them, where being holed up in a Michigan winter gave me the nerve to remake the manuscript in response to advice. Senior editor Doug Sery, who was definitely feeling overload, never lost sight of the big picture. Of the many experts at the Press who helped this project step by step along its way, Katie Helke Dokshina did so most consistently. Gone are the days when writers and editors can casually collaborate on the phone; the pace is now faster and software bloat has gotten in the way; but somehow the work still gets done, and very well. For a rigorous editorial reading of the proofs, I am much obliged to Caroline Constant, who had also helped very early in the project by providing short-term sabbatical digs in Boston.

Many artists and designers contributed images, not all of which I could include; some did so as much as two or three years before the project's completion. (All image credits appear with the figures. All photos and diagrams without any source indicated are my own.) For their outstanding work and kindness, my thanks go to Jan Edler of realities:united, Eve Mosher of HighWaterLine, James Sturm of the Center for Cartoon Studies, and to everyone involved at the MIT Press for the freedom to make the art selections work as they should. Thanks in particular to Yasuyo Iguchi, ever the master, for the book's design and cover art.

Special thanks to Kit and Cal, who always know who they are, each so much smarter than I, for their patience with my play-by-play descriptions of so many project stages. Even amid such preoccupations with work, they will always have my attention.

Last but not least, thanks in remembrance of Bill Mitchell, not only for his irreplaceable wit and wisdom, but also for blazing so many of the trails I have taken, and for keeping an early draft of my manuscript on his stack until the end, which came all too soon in June 2010.

"It's going to rain tonight."
"It's raining now," I said.
"The radio said tonight."

—Don DeLillo, *White Noise* (1984)

Prologue: Street Level

Out on the street, a cool night rain blurs the lights of the city, and water slowly drips off the signs. You step into a doorway to look at your phone. Kept dry there, up under the canvas awning, a speaker showers you with tinny tunes. Yet you no more notice these than the video display in the window, or the messages on your phone for that matter, for a splash from the sidewalk has soaked your shoes, and in this moment of disorientation, something just feels very good.

Maybe the water streaming over every surface helped, but, for a second, the world seemed of a piece, not just made of so many competing links, but somehow more immediate, with order and measure, a patchwork that for a moment you felt as one coherent space. In this sidewalk epiphany, the usual chaos of so many shallow short messages momentarily gave way to a presence that felt whole, of much higher resolution, in a word: replete.[1] What felt good was seeing more where you looked, instead of quickly wanting to look away. The rain brought out colors in the stone. The world became inexhaustibly detailed and present, in ways that a flickering picture is not. And, for a moment, maybe because you saw them streaked by the rain, even the glowing rectangles of your phone and the video display felt like features of this one, immediate, urban space, rather than simply portals onto other spaces, with furnished perspectives, all at no particular distance. For a moment, it seemed as if the sphere of information was embodied persistently in a physical commons. The sights and sounds of the city, the noisy, numbing vitality that leaves city dwellers experiencing daily life in a state of distraction, well, it all felt different for a second. Here was kind of attention that you could perhaps re-create, maintain, and manage, as if it would affect whatever else you notice. For one replete moment, you could understand the workings of attention, and how it might be worth knowing them better in an age of overload.

When you perceive the whole environment more and its individual signals less, when at least some of the information superabundance assumes embodied, inhabitable form, when your attention isn't being stolen, when you feel renewed sensibility to your surroundings you might try calling this *ambient.*

Ideas of the Ambient

I

Ambient 1

What do you notice? As a flood of information pours into ever more aspects of life, your focus becomes vital. Attention has become something to guard and to manage. *Ambient Commons* aims to help you cultivate yours through a rediscovery of your surroundings.

Right now your everyday environments are filling with ever more kinds of information, in ever newer formats of technology, used in ever more activities of life. Some of these make the world more understandable, even pleasant, but many less helpful ones prove difficult to escape. Whether carried about in your bag, hung on walls, or built into everyday objects, media feeds seem to be everywhere, as if people would suffer without them. Unlike the soot and din of a bygone industrial age, many of these feeds have been placed deliberately, and many of them appeal to the senses.

The appeal of this interface culture seems especially evident in the number of people walking (or driving) around staring at their smartphones. The interface arts have become the most prominent arts, especially since technology has spread beyond the desktop, work has left the office, and social play has networked at street level. There, as positional technology comes of age, new forms of interfaces reconnect to the world around—not just coordinates or tags for places to go, but also a dense aggregation of other technologies about environments, cities, and buildings.

Lately, that aggregation has been changing. There has been an invasion of glowing rectangles—ever more computer screens of ever more sizes, in ever more places.[1] And not just an invasion of screens but also one of networked objects, sensor fields, positional traces, information shadows, and "big data." This new era of interface designs is transforming the use of the city. Car and bike share systems for instance, would not have worked as well before now. Also on the rise are do-it-yourself applications and installations to monitor, tag, catalog, or curate everything from local plants to historical images to neighborhood lore. Many of these productions are said to "augment" their immediate surroundings, not just fill them with feeds and pointers to someplace else. Yet however much augmented, the city is also unmediated experience: fixed forms persist underneath all these augmentations and data flows, and for that you might be thankful. Without persistent environments, the sense of confusion and flux might only worsen. To have forgotten surroundings may indeed be a cause of overload in the first place.

Remembering can occur one detail at a time. A patch of sun slowly crossing a wall might produce a sense of calm while you work (figure 1.1). It might remind you how not all that informs has been encoded and sent. Its higher resolution and lower visual demands can restore your attention in a world that is otherwise too often low resolution and insistent.

1.1 High resolution, low pace: a patch of sun crosses a wall.

Similarly, the layout of your studio might affect the work you do there. Its tools and configurations could suggest possibilities. In many such everyday situations, the intrinsic structure of embodied space may affect your habits more than you know. It might actually shape some of your thoughts. Technology designers, neuroscientists, architects, and personal productivity consultants increasingly seem to agree on that.

Sometimes when people in many separate fields find themselves working in parallel on what proves to be one overarching issue, that issue acquires a new name. Getting a name then makes this subject matter more accessible, and brings still other fields into the conversation. "Light pollution," for instance, identifies an important interdisciplinary set of concerns that were formerly much more difficult to address. For many such phenomena, historians later take interest in how the name first caught on. "Information commons" arose in the late 1990s to emphasize intellectual property regimes on the net. "Ambient information" caught on in the first decade of the new millennium, as smart things, tangible interfaces, ubiquitous displays, and perpetual messaging all came into their own. "Ambient commons" isn't yet such a phrase: apart from a few musicians, almost nobody uses it. "Commons" itself seems familiar enough, maybe too much so through its many misuses. "Ambient" seems the more inviting way to begin.

What Is Ambient?

The word *ambient* usually occurs as an adjective. *Ambient light*, for example, lets you see the north face of a mountain, which (on the northern half of the planet) is always in the shade. Ambient temperature and lighting systems, the subject matter of environmental technology, are now being rediscovered by architects and put to use in green building.

Perhaps the most culturally resonant use of the word occurs as *ambient music*. Search "ambient commons" and you will find mostly musicians describing their small communities of practice. Ambient music is no small niche, however; critic Mark Prendergast proclaimed the last one hundred years of music the "Ambient Century."[2] His book by that name traced musical influences past landmarks such as Miles Davis's *In a Silent Way*, or more esoterically, the humming power lines of La Monte Young—all the way back to the pioneering works of Claude Debussy, who famously wrote: "I should prefer the creation of a music that has neither motifs nor themes, a more universal music."[3] Yet, across the intervening century, what really became universal was the listening, which became possible to do almost anywhere. Today you can buy underwater speakers for your swimming pool. If Debussy were to visit, how might he react to Pink Floyd cascading from supermarket ceilings?

Even more ubiquitous than audio, there is *ambient advertising*. This expression generally means advertising in proximity to the point of interest. It also means subtle: the advertising industry wants its messages to be remembered, but unnoticed. Subtlety has everything to do with placement. As one agency's mission statement once explained: "Ambient Planet delivers marketing messages to consumers wherever they may be and preferably in the environment where they would be most receptive to the message."[4] For a long time now, "out-of-home" advertising has been the biggest growth horizon for the industry. Although huge electronic billboards cause more controversy, smaller advertisements appear almost everywhere, as if no surface is too small to be left bare. Then, where every surface has already been covered, additional panels may be attached, as to the hoses of gas pumps. There is always room for more: as yet, nobody has bought the naming rights to the sides of the second base bag at Yankee Stadium, which has high exposure to camera feeds; even that could be unsurprising. Altogether, advertising stops at nothing. As a cultural force, it has few equals. And as environmental experience, it often leaves you little choice but to tune out the world.

Perhaps just as powerfully, social media also lay a claim to the word *ambient.* Small everyday signals of activity can be enough to tell you someone is around, to let you feel that you are together, without your having to speak or meet face-to-face every day. As explained in Lisa Reichelt's Twitter-friendly coinage of "ambient intimacy,"[5] social media use countless trivial messages to build a detailed portrait, even an imagined presence, of a friend. At least to some degree, this restores a lost kind of awareness found in traditional life. The upstairs shutters are opened, the bicycle is gone from its usual spot at the usual time, deliveries are being made, and the neighbors are gossiping. To their enthusiasts, social media re-create some of this environmental sense, albeit across the necessary distances and at the accelerated paces of the metropolis.

On the ascent of Twitter in 2008, *New York Times* columnist Clive Thompson explained the "paradox of ambient awareness":

> Each little update—each individual bit of social information—is insignificant on its own, even supremely mundane. But taken together, over time, the little snippets coalesce into a surprisingly sophisticated portrait of your friends' and

> family members' lives, like thousands of dots making a pointillist painting. This was never before possible, because in the real world, no friend would bother to call you up and detail the sandwiches she was eating. The ambient information becomes like "a type of E.S.P.," as Haley described it to me, an invisible dimension floating over everyday life.[6]

Ambient awareness can reflect a more general mindfulness, of course. Almost any use of the word *ambient* suggests some aspect of sensibility. Once considered irrelevant or a luxury in modern industrial cultures, sensibility to surroundings has become important again. In an era of changing planetary circumstances, personal attention to immediate surroundings seems like a manageable first step toward some huge cultural shift. Amid that transformation, the role of technology shifts as well, away from a means to overcome the world toward a means to understand it.

To information technologists, *ambient interface* represents an important new paradigm, with ubiquity and embodiment as first principles. Interaction design, the discipline best positioned to affect how you deal with technology, shapes not only sensory smartphones but also situated technologies. This is the form of ambient of most interest here.

Ambient Interface

The interface arts address the play of attention. To create ever more usable interfaces, designers work to reduce cognitive load: better design makes technology more intuitive and less obtrusive. Until recently, interaction designers have focused more on how users apply technology in the foreground of attention, as a deliberative task, for an intended purpose—and less on the role of context, or the importance of tacit knowledge, and how these shape intent. Within the arts of interface, ambient has been a fairly recent development.

Around the millennium, a paradigm shift from cyberspace to pervasive computing began to change those goals. Instead of mostly sitting passively at a desk, ("parking your atoms," as they said in the 1990s), users increasingly brought technology along into the existing world, in all its messy complexity, and sometimes they also built more of it in. Interaction designers turned to making new kinds of interfaces for this newly hybrid reality. The more their interface designs employed physical gestures or presence, became parts of larger objects, and seemed to fill the world, the more those interfaces could be said to embody information. To speak of *embodied information* helps to emphasize that not all computing is mobile; there are situated technologies, too.

Research in "ambient" technology began well before these recent interests. For example, one seminal project from the 1990s was the "Ambient Room" at the MIT Media Lab.[7] In this cubicle-as-monitor, the most noteworthy surface was the ceiling, which projected an animation of waves radiating across a water surface. The speed and size of these waves reflected different data of the occupant's choosing. This space may have been the first to move the display from figure to background. Since then, information technologies have moved the display still farther away, with nontextual background representations of markets, traffic, energy usage, and more. These technologies are no longer something to sit at; they no longer command full attention.

Ideas of the ambient have their philosophical basis in embodiment as described by mid-twentieth-century phenomenologists, most notably, Edmund Husserl and Maurice Merleau-Ponty. In his influential 2001 interpretation of these foundations for pervasive computing, interaction designer Paul Dourish explained: "I am using the term [*embodiment*] to capture a sense of phenomenological presence, the way that a variety of interactive phenomena arise from direct and engaged participation in the world."[8] Direct and engaged action makes use of innate familiarity with physical objects and situations. The early command line or later "graphical user interface" of desktop computers didn't allow for much of this intuition. Better future interfaces, better embodied in life's social actions, could eventually reward other kinds of attention. Many of these ideas came together at the famed think tank Xerox PARC, where anthropologist Lucy Suchman introduced Silicon Valley to "situated action," director John Seely Brown emphasized the attentional principle of "periphery," and chief technology officer Mark Weiser famously coined "ubiquitous computing."[9] To Weiser and Brown, periphery was everything you were aware of that didn't consume your attention, and that could be brought to the center of focus if necessary.

Much has happened in neuroscience since these ideas first caught on in interaction design. Today, ambiently embodied interfaces have active research communities, widespread instances in the electronic arts, and a burgeoning academic literature. Recent technological trends (which are the catalyst for but not the focus of this inquiry) have embedded computation into many more features of the world. Electronic positioning through the Global Positioning System (GPS), tagging through radio frequency identification (RFID), distributed environmental sensing (now cheap and wireless), and ad hoc local communication (Bluetooth) all contribute to these developments. Weatherproof displays on ever smaller and larger scales place live data into just about any setting. Augmented overlays for smartphones-as-cursors make the physical city browsable. Smarter, greener cities could become a "killer app" for these combined technologies. The expanding new field of *urban informatics* (also known as the *augmented city* or *urban computing*) seeks to collect, share, embed, and interpret urban infrastructural and environmental data. This agenda has become vital to the cultural imperatives of urban resilience, livability, and socialization. That makes interaction design ever more significant as a cultural endeavor, and more like architecture. It also makes it more ambient. As Dan Hill, an influential blogger of urban informatics once put it: "Multisensory interaction design now merges with architecture, planning, and urbanism, informed by a gentle ambient drizzle of everyday data—and so a new soft city is being created, alive once again to the touch of its citizens."[10]

Alas, pervasive computing can also mean ever more surveillance and ever more autonomous annoyances. The possibilities for overload and resentment seem endless. And alas, ignoring all these prospects may not work. To go back to a world without pervasive computing may be no more possible than going back to life before, say, electric lighting. So the challenge is to develop new cultural responses in the interface arts, as citizen sciences, and for the information commons. Although the latest techniques of interaction design lie outside this inquiry, they do help justify it. For as ever, any culture is better off seeking ways to make new technologies more aesthetic and more humane than assuming they will become so on their own, or wishing they would simply go away.

In A Single Name

Although *ambient* appears as an adjective in the increasingly familiar combinations above, it less often appears on its own as a noun. "Ambience," the most usual form, has less than helpful connotations, for it sounds like a luxury. It suggests a cultivated atmosphere that you might pay for in a restaurant, a theater, or a club.[11] By contrast, the design challenge of attention amid an age of information superabundance seems like a necessity to consider. This challenge invites a fresh form of use for *ambient.*

Thus this inquiry adopts *the ambient* as a simple name for a complex set of phenomena. This strange noun form is hardly intended to stick as a neologism, for it immediately seems far too inclusive for use anywhere but here. Here it may prove quite helpful, however. It usefully conflates several noteworthy conditions found where a new attitude about attention meets new era of information technology becoming situated in the world. You may never see or use this word in this way anywhere else, but for present purposes, consider twelve ways of describing the ambient (figure 1.2).

You may find some of these conditions more vivid than others. The ambient suggests some recognition of a whole, like noticing a forest and not just trees. Seen whole, most environments mostly seem unconsidered, chaotic, distracting, or forbidding. Exceptions exist in better buildings, neighborhoods, and cities—and in pristine wilderness, although today it is important to recognize environment as something more than places untouched by technology. Understanding the built environment

That which surrounds but does not distract . . .

Rampant availability of opportunities to shift attention . . .

A persistent layer of messages for somebody else . . .

No longer a luxury, but both a necessity and a nuisance . . .

An emergent effect of embodied interaction design . . .

An environment replete with non-things . . .

Semantic information, made more cognitively accessible . . .

Intrinsic (environmental) information, enlivened by mediation . . .

Cognitively unobtrusive media covering the cognitively restorative world . . .

A change in the nature of distraction . . .

A new sensibility and not the recovery of some lost sensuality . . .

A continuum of awareness and an awareness of continuum . . .

1.2 Twelve ways of understanding the ambient.

better has to be a step in understanding humanity's larger place in the world. Media situated in the built environment may help build that understanding. In comparison with any previous age, however, so many more facets of today's experience have been mediated so skillfully that they increasingly merge with perceptions of the world. The use of information technology has become increasingly circumstantial:[12] interspersed with other sensibilities, contingencies, and actions. Is this rich chaos of experience a new subject matter for design? Or for policy? Or environmental history? Does it become some new sort of cultural commons? This inquiry invites you to consider such questions. It has chosen to begin from the word *ambient*. If it had to

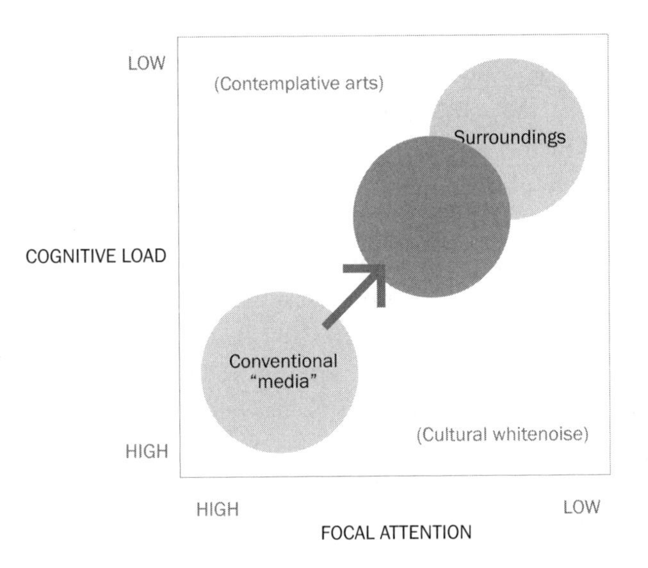

1.3 As media move into the background, do they augment it or just cover it?

begin from just one question, however, that might be this: do increasingly situated information technologies illuminate the world, or do they just eclipse it (figure 1.3)?

May the ambient invite tuning in instead of tuning out. May it do so with an emergent sense of a whole, or at least of continuum. Continuity seems lacking in a world full of separately conceived physical entities all competing for space and attention, all without concern for what is nearby, and masked by portals, links, and signs to someplace else.

The word *ambient* has its origins in embodiment as a cosmological embrace—the firmament seen as a protective dome, as the vault of the heavens. In the ancient Greek worldview, under the aegis of the gods, the sheltering sky was inseparable from what the modern age calls "air."[13] To say that an idea has been "in the air" is quite normal and reflects a sense that each place and time has its own intellectual atmosphere.

Today, the rediscovery of air has become a prominent philosophical theme. After all, atmospheric science tells the most profound story of the times. Thus it has become a popular parable to retell how, in the eighteenth century, Joseph Priestley discovered oxygen—how it is oxygen specifically, and not some abstract divine *aether*, that one must inhale for life, and that fire needs for combustion.[14] Earlier, René Descartes knew air only as "matière subtile"; Newton referred to it as an "ambient medium"; earlier still, Galileo Galilei referred to physical phenomena being contained in "l'ambiente."[15] In more ancient use, the word *ambient* goes back to the Latin *ambire*, to "go round," "visit in rotation," "inspect," "solicit," "canvass," "encircle," or "embrace." The idiom "aer ambiens" meant "atmosphere," the "encircling air," although not quite the "sheltering sky."

This legacy of the ambient appears most richly probed in a rediscovered essay, "Milieu and Ambiance" (1942), by the Viennese philologist Leo Spitzer.[16] *Milieu,* which combines the prefix for "middle" with the word for "place" in French, was the more widely used word at the time, in English as well, and meant a person's social setting, surroundings, or ensemble of circumstances. Like the German *Umwelt* (literally, "surrounding world"), *milieu* and *ambiance* also implied an environmental influence on outlook or action. As such, both words often served the purposes of literary realists and sociological positivists of the late nineteenth and early twentieth century, the influential Berlin sociologist Georg Simmel for one, to whom the modernist notion of distraction is most normally traced. In "Milieu and Ambiance," Spitzer connected this modern sociological notion back to ancient Greek and Roman cosmologies in which mortals did not merely occupy their environment, but indeed were embraced by it. "For the Latin verb *ambire* had not only the literal meaning of the Greek verb περιέχειυ; it possessed as well the same connotation of protection, of a warm embrace: cf. [Pliny]: "*domis ambiri vitium palmitibus ac sequacibus loris,*"[17] Although competing versions and translations of Pliny the Elder exist, one reading is this, "a country house embraced in the shoots of closely reined straps of grapevine."[18] Except now make that carefully switched runs of fiber-optic cables. Today, that embrace is by information.

A New Firmament

For a metaphor of this embrace, imagine information technology as a modern firmament. The famous woodcut from Camille Flammarion's *The Atmosphere* (1888) depicts a firmament in

Un missionnaire du moyen âge raconte qu'il avait trouvé le point
où le ciel et la Terre se touchent...

1.4 Firmament and beyond: the "Flammarion woodcut" (by an unknown artist that first appeared in Camille Flammarion's *L'atmosphère: météorologie populaire*, 1888) (Wikimedia Commons).

which a medieval man dares to look beyond the sheltering sky, into the modern technological world (figure 1.4).

When a lonely person pulls out his or her smartphone, is that to seek the cosmological comfort of a new kind of firmament? To judge by the widespread casual use of media, the very presence of information seems to soothe more than its absence. Restaurants place televisions around as much for comfort as for any actual viewing. Stores all play music, and if you ask why, their managers will tell you that silence is bad for business, for it makes people uncomfortable.

Notions of media dependencies amid jaded overload have been prominent in cultural theory for more than a century, ever since the work of Simmel and his contemporary, Emile Durkheim, became known. Indeed, ever since modern industrial urbanization, when so many new technologies pulled so many communications out of their traditional cultural contexts, a blasé outlook has been the norm among city dwellers. The time has come to reconsider that oft-cited trope. Today, you must remember to look to the other side of the technological firmament, into a timeless but endangered world of direct experience.

The experience of technology has changed so much since the industrial city and the heyday of print and broadcast media that it is time to reexamine the urban citizen's distraction. Much of the change has been away from command by any one information medium. The world has been filling with many new kinds of ambient interfaces. Nothing may be designed on the assumption that it will be noticed. Many more things must be designed and used with the ambient in mind. Under these circumstances, you might want to rethink attention.

1.	AMBIENT
Main idea:	The ambient could illuminate the world or just hide it
Counterargument:	Who cares about context?
Key terms:	Ambient, interface, firmament
What has changed:	More contexts and formats of information
Catalyst:	Pervasive computing
Related field:	Interaction design
Open debate:	Holding onto reality?

Information 2

Light in the east means day is coming. Although a sunrise can mean much more, such as a fresh start, or hope, it always indicates a day, and it does so without intervention of language, symbolism, or code—no instructions are necessary.[1]

At a crosswalk, a sound signal slowly ticks until the lights change and pedestrians may safely walk. Then it rapidly tocks. This audible code has been installed not just for the blind, but also for those whose attention is somewhere else, fixed on a friend or a phone perhaps. Any set of sounds might work for this purpose, but higher frequencies would carry into neighboring buildings, and lower frequencies would get masked by the rumble of trucks going by. A synthetic voice could command crossers to "Wait! . . . Wait!" but that would be unnecessarily disruptive.[2] A better design would provide a signal clearly audible to those waiting to cross but not audible to others who aren't at the crosswalk, one that allowed waiting pedestrians to carry on with other activities, like conversation or texting. For even those who don't push the crosswalk signal's activation button are subjected to its effects and become involuntary "users" of the experience.

Five friends work their way across the metropolis on crowded transit lines. Along the way, they text one another to negotiate a destination. The outcome will depend on positions of other nearby friends sitting in cafés, where they have checked in using social navigation apps such as FourSquare. The five use the interplay of noise and information to drift gracefully from group to group.

One morning, as you arrive at work, you notice a brand-new sign declaring that parking is to the left. Was someone unable to see the cars parked over there? Perhaps there is someone so steeped in perpetual messages that nothing remains self-evident. Perhaps everything needs applied documentation, as when, to learn another language, you go around the house putting yellow stickies on everything, with strange new names for things you know well.

These examples illustrate a spectrum of mediation. Although each describes a situation that informs, they differ in the degree of instruction, from none for dawn, to adequate and impersonal at the crosswalk, to abundant and personal between texting friends, to superfluous and impersonal at the parking lot. They illustrate how just enough mediation can be helpful without intruding, how even ample mediation can be both helpful and pleasant, but how superfluous mediation reinforces being out of touch with the world. As explained by philosopher Albert Borgmann, some information is about reality; some is for reality, to create it, as plans do; and some gets taken *as* reality, that is, without referents in unmediated experience.[3]

They also illustrate augmentation. Richer, more enjoyable, more empowering, more ubiquitous media become much more difficult to separate from spatial experience. You perceive the world largely through what you can do with it. Today those possibilities have an ever broader spectrum and deeper volume of mediation. Latter-day milieus are mostly found in communications. Having noted the origins of the ambient in notions of cosmological embrace, you might now ask if today's embrace is by information.

Superabundance

In ways perhaps second only to planetary change, information superabundance is the quality that most distinguishes these times from even very recent ones. Unlike planetary change, superabundance gives cause to rejoice: all the world at your fingertips, on demand, and no need to clutter your head with it. Having a hundred times more items available on any given topic than there were just a short time ago might seem daunting, if they weren't so often a thousand times easier to find, filter, sort, and trace. You hardly meet anyone who laments how much better life was before Google. Young thinkers today might well wonder how, before the net, anyone pursued ideas that interested them.

Just how much information has become possible to estimate. According to a famous study in 2003, for instance, the amount of new information that year was the equivalent of forty-seven times the contents of the Library of Congress.[4] In the decade since, the annual flow rates of many individual genres, such as video sharing, phone call archiving, or library-like reference materials, may each have reached that volume. In 2010, the

Economist reported that humankind first produced data at a rate of a trillion gigabytes per year in 2009, up by factor of 8 in half a decade.[5] Also in 2010, Google CEO Eric Schmidt announced that "every two days we create as much information as we did from the dawn of civilization until 2003."[6]

Information ethicists caution that, by overwhelming your capacity for reflection and empathy, this superabundance poses a real threat to judgment, well-being, and relations with others.[7] Being overwhelmed with information tempts people to accept what is most agreeable and convenient. Convenience discourages extra expenditure of attention. Not everyone takes the trouble to judge the quality of the information items they most easily find, to reason about opposing views, or to remember that it is only information of a kind, and not the world itself.

Choice theorists caution that, beyond some sufficient threshold, abundance just bewilders. The more kinds of soap on the shelves of your supermarket, the less likely you are to buy any of them.[8] Whether choosing a hotel, reading about the music you are listening to, or debating a finer philosophical point, any moment of curiosity can quickly produce so much information that you may feel helpless. Having so many options only increases the chance that other aims will arise before you can act on your present ones, all the more so when your inquiry brings forth information with weak connections to your present aims. Thus, as is commonly known among designers, but suffered everywhere anyway, data that do not inform only produce anxiety.

Social historians often warn of the unintended consequences of sudden infatuations with new technologies. Just as Americans rushed to do anything and everything in cars half a century ago, so, today, people worldwide are rushing to do anything and everything on socially linked smart devices, often all at once. It was decades before experts recognized the physical, social, and environmental health consequences of overreliance on the automobile. How long will it take to recognize the consequences of much wider overreliance on smart devices?

All told, today's flood of information seems far less than the absolute good technological stakeholders make it out to be. To an educated person, retained knowledge would have once been much more of an asset. To just about anyone, life had fewer occasions of bewildering detail and more to use common sense. An older person who spent more time outdoors growing up than generations since might marvel how youngsters walk around with headphones but without coats, connected but oblivious, bare-toed in flip-flops in the snow. So called "digital natives" counter that the outdoors is a bore, and this superabundance of information is almost all delight and no distraction.[9] Whatever you grow up doing reshapes your brain, they claim. And however you have educated yourself, you have learned how to tune most things out, even freezing toes.

Today, at one and the same time, there exist some of the most distracted human beings ever alongside unprecedented new degrees of attention management.[10] For the sake of argument, consider such management as involving two basic approaches to superabundance. One consists of good filtering. Paradoxically, the best way to deal with so much information is usually more information: metadata, usage histories, recommendations, source certifications, blockers and aggregators, social networks. By contrast, a second approach consists of mindful abstinence. Of choosing to ignore noises from the street; to skip

24/7 news; to turn off the smartphone when with friends; to multitask only for fun; to stay off the net one day a week; to ignore the latest titles from Hollywood; to listen to live music often; to look up at the architecture around you from time to time, and not just down at your screen; to enjoy something without feeling the urge to message a friend about it; to know what phase the moon is in, and how to stay still long enough to watch the stars wheel across the night sky, at least where you can actually see them. Today both of these approaches seem necessary, to some degree. What is more, attention to surroundings may serve both of them. To abstain from some less necessary stimuli may help you notice the role of context for filtering other more necessary ones.

Today almost anyone can admit some sense of information overload, most often describing it with metaphors of flow: drinking from a fire hose, surfing on chaos, streaming your feeds. Thus even the labels on bottled water sold at The Henry Ford, a museum with millions of artifacts, bear such a metaphor: "In your quest for knowledge, be sure not to drown in all the information."

"Information overload" was coined by *Future Shock* author Alvin Toffler about forty years ago, and "information anxiety" by Richard Saul Wurman about twenty-five years ago. (Figure 2.1 graphs the usage of *overload* since 1910.)[11] Despite coming before the web, search engines, or smartphones, many of Wurman's conclusions remain pertinent today. "The word 'inform' has been stripped out of the noun 'information,' and the form or structure has disappeared from the verb 'inform,'" he cautioned. To be informative, data require not only proper design and form-giving, which was Wurman's expertise, but also at least

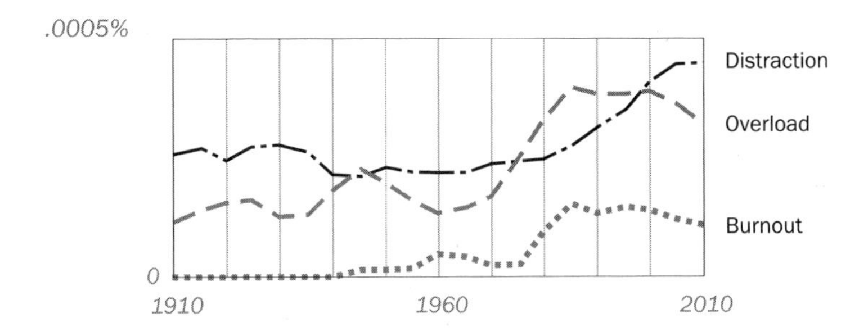

2.1 Increasing use of *overload* and related expressions (a Google nGram).

some degree of attention. Overload has to do with capacity, and that has to do with sensibilities. "Before you can curb your information intake," Wurman advised, "you have to come to an understanding of who you are."[12]

"You are what you attend to,"[13] to paraphrase the first modern psychologist, William James, whose work from over a century ago remains an obligatory part of the attention literature to this day. Now, as a paradigm shift in cognitive science has many asking if anyone really knows what attention is, this inquiry will need to go beneath the surface of such aphorisms.

The problem of attention was also famously identified by the information scientist Herb Simon, in another inexhaustible aphorism, written right about the same time that Toffler coined "future shock." Simon, the main thinker behind the rise of Carnegie Mellon, wrote in 1971: "In an information-rich world, the wealth of information means a dearth of something else: a scarcity of whatever it is that information consumes. What information consumes is rather obvious: it consumes the attention of its recipients.

Hence a wealth of information creates a poverty of attention and a need to allocate that attention efficiently among the overabundance of information sources that might consume it."[14]

Overconsumption

"Is there anywhere on earth exempt from these swarms of new books?" asked Erasmus, the first modern editor, in the sixteenth century.[15] Indeed, sensitive souls through the ages have complained of information overload, long before there were computer networks.[16] The historian Ann Blair has assembled many such remarks from scholars in information overload's early years. "The early modern experience of overload was different in many ways from today's," Blair has observed. "For example, then only an educated elite and a few areas of life were affected. Today people in nearly every walk of life, at least in the developed world, rely on the Internet for much of their basic information. What we share with our ancestors, though, is the sense of excess."[17]

Because overload is a perception of excess, anyone in any era might have felt it, wherever circumstances exceeded expectations.[18] Historians often speculate whether, in a world where a printed page or painted image was rare, it might have had a much greater impact. Or whether, in a world not yet saturated in media or disenchanted by modern science and economics, direct experience itself might have been much more vivid, perhaps overwhelmingly so.[19] Whenever a new medium arose or become arbitrarily fashionable, it may have been experienced entirely out of proportion with existing cultural capacities to make sense of it. Thus a seventeenth-century Amsterdam merchant gathering news of arriving ships might well have felt overwhelmed with information, though he wouldn't have used that word. Nineteenth-century commentators ranted about the growing obsession with that antisocial new creation, the newspaper.[20] In a world where mediated communications were not nearly so pervasive as they are today, such overconsumption was more localized, confined largely to the literary salon. Besides being confined to particular settings, this overconsumption was also confined to a privileged few. Some literary critics think the first person to suffer overload was a nobleman: Don Quixote, the fictional protagonist of the first modern novel, became delusional from overconsumption of medieval romances, formerly scarce but now easily obtainable in print editions.

"Just as fat has replaced starvation as the nation's number one dietary concern," sociologist David Shenk wrote in the 1997, "information overload has replaced information scarcity as an important new emotional, social, and political problem."[21] At least for the luckiest billion humans, the supply of mediated stimulus, like the supply of food, is at an all-time high. When a formerly scarce resource becomes abundant, people instinctively consume it wherever and whenever they can, which naturally leads to overconsumption. Empty calories in the echo chambers of blogs and retweets instantly gratify the need to tell somebody whenever anything occurs to you. Much as with overconsumption of unprecedentedly available sugar and fat, overconsumption of informational empty calories can lead to what some critics now call "information obesity."[22] Messages, music, movies, and market prices are always on and gently drift by, not just at the workplace but in most settings. Habits of consuming and switching among these make new sorts of interruption seem acceptable. When you can so easily search for a fact or identity in just a few seconds, sometimes you might interrupt a conversation to do so (hopefully not with the very person you are researching). Smartphones make messages available anyplace, amid any activity. It is not so unusual, for example, to hear a business deal being negotiated from the next toilet stall. Overconsumption comes not only from an increasing volume of communications, but also from an increasing disregard for context.

Information, of a Kind

According to its historians, the word *information* did not come into general use until the nineteenth century, when the rise of industrial technologies created massive amounts of printed matter to organize and store, and so led to libraries, catalogs, mass-produced newspapers, and disembodied telegraph signaling.[23] And the more the world became raw material to be exploited, the less it seemed to inform as it was. Information became a thing to be stored and retrieved, and not just the process of gaining knowledge from data in a situation.[24]

Only with electronic communications did the word *information* begin to correlate with mathematical encodings, such as the ones used by Western Union's telegraph transmitters. With the rise of cybernetics in the mid-twentieth century, the mathematical theory that has since been the dominant meaning of the word took form. Today, the word conveniently comprises both the data packets that travel down optical fibers and over the airwaves and the use that someone makes of those data packets.

To question the prevailing meaning of *information* quickly becomes an exercise in epistemology. Thus a widely cited argument by Michael Buckland holds that information can be a subjective process, an aspect of knowledge, or a thing (i.e., capable of being processed technologically): "Being 'informative' is situational and it would be rash to state of *any* thing that it might not be informative, hence information, in some conceivable situation."[25] After all, at least some knowledge comes from direct experience. Only modern communications technology created an emphasis on sent packets of data. The point here is that the balance has shifted away from tacit knowledge through direct experience toward explicit but indirect knowledge through communications. French makes this epistemological distinction in its verbs *connaître* (to know personally through acquaintance) and *savoir* (to know of, through learning.)

According to the widely respected information ethicist and semanticist Luciano Floridi, the prevailing mathematical theories of information were really theories of data communication, that is, "uninterpreted symbols in well-formed strings of signals." Floridi has advanced a more carefully qualified definition of *information*: "true semantic content."[26] With certain exceptions (such as sunrises), being informed generally involves linguistically encoded meanings, at appropriate levels of abstraction, all made intelligible by frames of knowledge. Without these, transmissions that are meaningless or false may too easily be taken for information.

Once you assume that most information is semantic, you can more easily identify its opposite—nonsemantic information. For one self-evident example, light in the east informs without symbols. The dawn need not be a social construction. Philosophers call this assumption of information without inscription the "realist thesis."

For example, bear tracks on a wilderness trail indicate that at least one bear has passed that way. A picture of these tracks

2.2 Intrinsic structure informs. The rings of a tree stump show its age. Photo: thousandShpz/FlickR.

would be a document that you could store and retrieve, but the tracks themselves aren't a document. Although they refer to something not present, they do so without an agreed protocol or intended communication, and without human agency. A raccoon could also tell that a bear had been by. Such nonsemantic information takes no language or encoding.[27] Linguists call it "natural meaning."[28]

Floridi calls nonsemantic information like animal tracks or sunrises "immanent data," where "immanent" means that the form or structure of some phenomenon is coupled to the state of another, so that form of one indicates state of the other, as fingerprints left at the scene of a theft indicate who were the thieves. Floridi believes such immanent data should be understood as "environmental information."[29] In a vivid textbook example of environmental information, the rings of a tree stump reveal the tree's age when it was cut down (figure 2.2). Because the word *environmental* has so many other meanings, however, and a specific, different one later on here, this inquiry calls nonsemantic information such as fingerprints and tree rings "intrinsic information" instead.

Intrinsic Structure

"Intrinsic" means "inherent to the material, structure, or constitution of something." *Intrinsic information* exists mainly on the scale and in the form and configuration of learned surroundings. To speak of "embodied information" is to address the combination of mediated with intrinsic information (figure 2.3a).

Especially through their embodiment, persistent structures and recurring patterns provide the frameworks by which to assimilate new or changing phenomena. For example, when walking down a city block, you might notice that one house has a new fence, but only if you have walked that block often enough before. A notice on the fence might warn you to keep out, but the structure of the fence did so already, in a less mediated way.

With intrinsic information, content remains inseparable from form. The effect of form, though not always a sign, can be experienced nonsemantically, through its embodiment. When a bike rack has been placed outside a café, you don't need instructional signage to park your bike there, nor a "This side up" label to remind you how to park it. These are further examples of "natural meaning."

As new forms of packaged information arise, in ways that reduce cognitive load and that appeal to the senses, it seems important to uphold the importance of the intrinsic structure, and to find fascination with it. The unmediated world would

(a)

Semantic

Intrinsic
(environmental)

Transmitted Situated

Symbolic
interface

Tangible
interface

Direct, without
intended interface

Mediated Unmediated

(b)

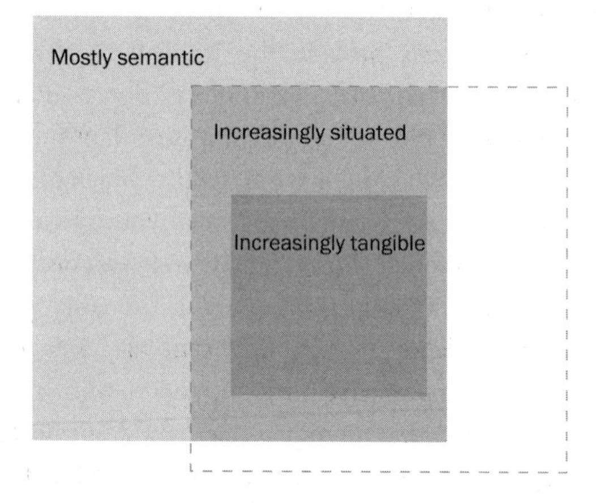

2.3 Loss of intrinsic information. (a) The ambient increases situated, mediated, and tangible kinds of information. (b) In an imaginable limit case, to be avoided, the unmediated, direct, and intrinsic kinds of information have been forgotten.

suffice as the ambient, but only to those whose culture provided ample experience with intrinsic information, as reflected in street sense or traditional folklore. Conversely, a totally mediated world, which seems to be where much urban space is headed, would also suffice as the ambient, to those with sufficiently fluent access to the pervading media. So the imaginable limit case of the ambient would exist without intrinsic information (figure 2.3b).

To put it more plainly, intrinsic information seems increasingly at risk. Pervasive media could assume its form as a double, but in doing so might mask it. Technologically dependent sensibilities become less inclined to notice much else. To media-addled kids, the unmediated world is boring. They simply haven't developed the cognitive capacity to appreciate it.

Traditional cultures' default emphasis on intrinsic information waned long before the rise of computer networks. Practical means of transportation and the written word were enough to tip the balance. The first guidebooks and street signs of the early modern city provide a noteworthy chapter in the history of information. Print and electrification also augmented the city, but not so pervasively or seductively as information technologies now do. Never before have so many inscriptions so literally and figuratively covered so much of the city; never before have inscriptions been quite so cognitively engineered. Never before have so many interfaces to otherwise unrelated processes been so similar. As a result, many everyday processes of reading, navigating, shopping, and socializing resemble one another more now than they did a century ago, and much more than they did in traditional cultures. As an unintended consequence, these processes rely somewhat less on intrinsic information.[30]

As another way to understand such detachment, consider intrinsic information in relation to the abundance of textual signs. When modernity had filled so much of the world with organized communications, unsurprisingly this led some philosophers to propose that everything communicates. Or, more specifically, to propose that the city as a built environment becomes a text, at least of a sort. Starting with the industrial age, cities not only labeled but also increasingly shaped and organized themselves according to deliberate schemas, such as street numbering, visitor guidebooks, corporate identities, advertised attractions and so on.[31] From this industrialized experience came theories of attention, systematized communications, and signs. To the American pragmatist Charles Sanders Peirce, for example, the "semeiotic" (as he spelled it) embraced most other categories of his philosophy. A sign was anything some party devised to represent something; moreover, it was nothing without interpretation. This approach led toward a more general theory of signifiers, generally credited to the early twentieth-century linguist Fernand de Saussure, who explained how the designation and interpretation of signifiers could be arbitrary, that is, not necessarily bound to intrinsic features, or what we are calling "intrinsic information." Rereading Saussure, late-twentieth-century critics saw free play of those signifiers in increasingly constructed social interpretations. In the process, especially after Roland Barthes in the 1970s, *semiotics* became a catchword both for the situational and unintended aspects of cultural communications and for their critique. To this way of thinking, buildings themselves were mere signs, and the city was surely a text.[32] But in the process, and perhaps due to cognitive theories of symbolic processing that were emanating from computer science at the time, the unconstructed, unmediated aspects of environments were forgotten.

Long lost by then were Peirce's qualifications that not all aspects of an object signified, and that not all aspects of sign production referred to things. Underneath the free play of signifiers, an unmediated world was still there, one that worked without need for encoded transmissions—and it wasn't a text.

The Eloquence of the World

This chapter began with an everyday example of superfluous signage, as if nobody could recognize a parking lot without instruction. That was one small instance of the consequences of modern information superabundance. But what if information technology becomes so complete that nothing remains to be discovered and documented? [33] Or, as the novelist José Luis Borges asked over thirty years ago, what if the map grows larger than the territory?[34] When we know almost everything through documentation, and almost nothing directly, will any of us notice that something is missing? And when we know everything only through its signs, what will happen when the signifiers start playing more freely and their referents no longer refer to the things they once did? And, finally, what if the signifiers become reality itself, much as an MP3 stream is casually taken for the music itself, right there in your earpods, and not a recording of a musicians' session somewhere else?[35]

Many philosophers have voiced this challenge as one of balance. Thus, for Albert Borgmann, "Righting the balance of information and reality is the crucial task. It amounts to the restoration of eminent natural information."[36] There is something to be lost. "The task, it would seem, is to commensurat[e] the fluidity of information technology with the stability of the things and practices that have served us well and we continue to depend on for the material and spiritual well-being—the grandeur of nature, the splendor of cities, the competence of work, fidelity to loved ones, and devotion to art or religion. What is needed is a sense for the liabilities of technical information, and an ear for the changing voices of traditional reality."[37] Floods of encoded information can drown out what Borgmann called "the unsurpassable eloquence of reality."[38]

Knowledge of the world makes life more sane and rewarding. Knowledge from acquaintance with enduring physical conditions provides a framework for other, more ephemeral ways of knowing. In language, for example, an intuition for the spatial relationships of physical objects grounds much metaphor, and metaphor enables much thought. In masterful work, known relationships of tools, media, and processes allow creative responses to unforeseen situations where these relationships are present.

Information technology has clouded these principles because mediated cultural productions now appear with such high resolution, in so many different circumstances, that many more people take them as reality. The cultural experiences of video, navigation, data simulations, and social networks occur through increasingly rich media, under ever easier suspension of disbelief. The immediate gratifications of multitasking among such media tend to mask or preempt the slower mental states of appreciating the unmediated world. Many more people have grown up without having gone outside to play as children, or without ever having chosen to get lost in a city as adults. Some send email instead of walking down the corridor to talk to a colleague. As educators often lament, some of their students can't always tell the difference between fact and fiction, sample and simulation, or visualization and appearance. Some people still buy SUVs as surrogates for the cold, messy experience of ever visiting actual mountains. Some have spent more time in nature stores than outdoors.

"Information as reality" is how Borgmann put it. "Information can illuminate, transform, or displace reality."[39] Now comes increased ethical challenge to keep these in balance. Information "about" reality describes and reveals. Mythology, poetry, and geography also refer to the world as it is. Information "for" reality makes bold plans, and realizes them through the contingencies of the physical world.[40] Contingency seems vital to Borgmann's argument, which makes its case through architecture. Thus the magnificence of a late medieval church comes not from having been conceived and made so all of a piece, but instead from having been accreted and adapted over centuries, in different styles and constructions, all toward a persistent vision. "When the creative power of humans and the contingency of reality are consummated in a great work, the latter has a presence more commanding or expressive than that of a text, score, or plan."[41] The Latin root *contingere* connotes consummation more than circumstance. The enduring, situated doneness of the work gives it its power, especially for orientation and memory. Without contingency or permanence, and without enough difference between things that are made and signs that prescribe them, information would not be "for" reality. Lacking these epistemological anchors, information "as" reality can be anything one moment, and something else the next.

To those who equate civilization with some sort of permanence, and to those who lament the loss of environmental knowledge, this may suggest a new dark age.[42] Environmental advocate Bill McKibben once proclaimed "the age of missing information."[43] Writing twenty years ago, long before YouTube or Google, McKibben claimed to learn more from a day on a mountain in the Adirondacks than from watching a day's worth of television, which his friends recorded for him while he was out hiking. From an epistemological standpoint, it appears that McKibben was enjoying information about the world through the natural meaning of things, without encoding or human agency. When he called this "information," that was in contrast to the usual technological sense of the word as something sent.

Alas, in many disciplines of thought, a fundamental break has occurred with the intrinsic structure of the world. "If we allow having information to obscure the difference between direct and indirect knowledge," Borgmann cautioned, "our sense of the presence of things will become obtuse."[44] Whether in finance, art, or neuroscience, mediation has a presence of its own, and symbolic processing too often turns away from the external referents of its operands toward some internal elegance of its algorithms. As a result, semantic processing has become much more important to cognition research than it was half a century ago, when primary perception was the main topic. Neurolinguistic information processing models increasingly dominate the discipline. In many ways this has made symbolic context more important than physical context. And because seated, media-based tasks prove easier to test in controlled experiments than embodied environmental engagement tasks may ever be, it has led to far more research along these lines, with the focus turned from embodied social behavior (like the perception of crowding) to divided personal attention. When it seems as if everyone is staring at glowing rectangles, you tend to experiment using those rectangles. Here, too, something may have gone missing. If humans have a propensity for tacit environmental knowledge, the cognitive neuroscience disciplines ought to find more to say about that.[45]

In sum, the ambient raises a knowledge challenge. As media become peripheral and less cognitively demanding, do they become more environmental? Do ambient media eclipse or enhance awareness of intrinsic phenomena? Somehow the role of tacit knowledge of embodied space has been underestimated. How that knowledge becomes the basis for larger, more abstract mental models of the world is just one aspect of the problem. There is a more basic disregard for settings, which only feeds the belief that "the environment" is somewhere else and hard to do anything about. When almost everybody is disconnected, almost nobody will notice that something is missing. It must be an epistemological oversimplification to say this, but people don't even know that they don't know, say, for whom the streets have been named, the name of any tributary to their local river, which vegetation is native, the prevailing wind directions, or what phase the moon is in. And, on the rare occasion of a lunar eclipse, you could suggest to a digital native that this might really be worth checking out, only to get a response of "Oh . . . , cool . . . , where?"[46]

2.	INFORMATION
Main idea:	Not all information is sent; the world itself also informs
Counterargument:	Semantic theories of knowledge
Key terms:	Overconsumption, intrinsic information
What has changed:	Definitions of information
Catalyst:	Superabundance
Related field:	Epistemology
Open debate:	How to put value on intrinsic information?

Attention 3

As attention grows scarce in this age of superabundant information, now might be an excellent time to learn more about it.[1]

You might first think of attention as a kind of focus. Even if you turn off all your screens, focus is hard to sustain. Thoughts dance about like a flame. Words turn into language games. Vision naturally drifts, as a means of alertness. Incidental properties such as texture stand in for more essential ones, and they help memory make lateral associations. Perception of course lets only a few stimuli through, and many of those automatically. Not all attention involves thought.

You may have learned as a child that attention was something to pay. Later on, you also learned how attention could flow, or be divided, or be carefully conserved. Business practices such as slide shows assume that you aren't going to pay attention unless you have to. Recreations restore attention through flow. Entertainments divert and divide attention, often by triggering deeper, more automatic responses, to bright moving objects, say, or to sudden loud noises or slow zoom-in shots of food. Thus, besides being something to pay, attention can be involuntarily stolen, consciously fragmented, or effortlessly restored. So you might take issue with James's aphorism "Everyone knows what attention is."[2] You might wonder just what it is that can be divided so acceptably, appealed to so relentlessly, or dissociated from physical context so universally by today's communication media (figure 3.1).

3.1 Divided attention at street level. Cartoon by James Sturm.

Divided Attention

Although the human mind naturally tends to wander, never before has it had such abundant means for doing so.[3] This is an age of unprecedented distraction.

As a habit of mind, distraction may not have changed much over the ages. To reach it still takes no more technology than a jug of wine or a pair of dice. But as a cultural project, and a major industry, distraction seems much more recent. Although any culture had its entertainments, none before had so many technologies for filling so many gaps in time or space with them. Whenever tyrants and theocrats had sought total control of communications in a society, the scope of their efforts was limited. Nothing has succeeded quite like modern consumerism at saturating life with its messages.

Today it seems difficult to imagine that, until the last century, most settings of everyday life remained beyond the reach of communication media most of the time. Even the most actively engaged citizens spent much of their day away from media. Modernity was in essence a shift beyond that; and of all modern technological advances, electronic communication may have brought the most change. Electrification introduced many more frequent forms of connection with infrastructures. It created new technical delights, indeed a new discipline of industrial design, and brought many of these into the home (figure 3.2).

It helps to remember that only with the advent of the Internet did two-way communication become inexpensive enough to be always on, however. Only then did communication become asynchronous, many to many, visually modifiable, attachment laden. And only in the last decade did Internet connectivity get worn, carried about, and embedded into things not normally

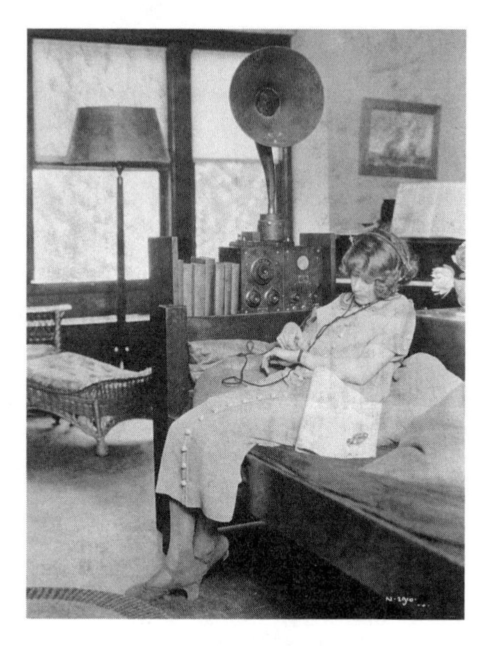

3.2 The modern increase of technological interfaces in the home: setting a watch by radio, 1925. Photo: Phyllis Frederick (Schlesinger Library, Harvard).

perceived as communication devices. And embedded into life: today, just about anything human beings do has been done while texting.[4]

Of course, the appealing ease of today's information technology encourages multitasking. Formerly quite distinct activities such as work, play, and shopping now not only appear on the same device, but also share the use of similar symbolic manipulators. In using these, many operations can become automatic enough.[5] Skills acquired from one kind of task transfer and indeed lead to others. This creates pleasure in itself, from just how many kinds of things you can do with a few acquired interface skills. Full of useful tools, instantaneous references, personal configurations, and curated finds, the smartphone itself has become an object of status and desire, to be used for the sake of its usability.

Many people associate divided attention with the workplace. Among the contexts of technology use, the workplace most inherently provides interruptions, interrupted by other interruptions—many not by coworkers but self-initiated. At work, multitasking lets you coordinate concurrent activities, research facts on the fly, switch tasks when you are no longer fresh on the current one, and feel the comfort of moving things in and out of focal attention.

At home, divided attention lets you relax with as many of your personal media selections as you care to enjoy at once or in alternation. It seems almost a cancellation process: a retreat from an annoying set of glowing rectangles at work to a soothing set of glowing rectangles at home. There might be advantages to unplugging altogether for a while, but people seem to put a higher priority on regaining control of their media first.

Under these diversifying technological circumstances, attention moves about more readily. It is easy, for example, to interleave reading a book electronically with chat, social networks, shopping, a movie, grazing at news feeds, or perhaps writing something of your own. There can be pleasure in mixing activities when some are physical and others mental but not too challenging, like listening to music while running. Concentration can be difficult, and multitasking can be a form of relief.

Increased social connectivity brings increased "fear of missing out" (FOMO), which, in a self-propagating loop, leads to ever more messages of ever less substance. As eminent social psychologist Sherry Turkle has cautioned, connectivity can reach the point where people mostly opt for communication formats (such as texting) that allow them to divide or defer attention, even to friends, even those present.[6] To monitor so many feeds so continuously takes what design strategist Linda Stone has named "continuous partial attention,"[7] a wider, more reactive alertness than multitasking, which emphasizes only a few concurrent actions. Though not solely social, such attention is motivated by FOMO. People do less in order to remain aware of more. "In the case of continuous partial attention, we're motivated by a desire not to miss anything. There's a kind of vigilance that is not characteristic of multi-tasking," Stone has explained. Usually in everyday multitasking, at least one of the activities being multitasked is "somewhat automatic, like eating lunch or stirring soup. That activity can be paired with another activity that's automatic with an activity that requires more cognitive resources, like writing an email or talking on the phone."[8] Continuous partial attention is instead the desire to monitor many things at once, in case one of them should do something interesting.

Naturally, there are limits to how many distinct things you can keep your focus on. It doesn't take a neuroscience lab to demonstrate this. A simple exhibit at San Francisco's famed Exploratorium shows just how few. It will ask you to watch a blue square among some dozen green ones as they all move about the screen, until the blue square turns green and seems to disappear among the other green squares. Then, a few seconds later, it will ask you to identify which of the squares was once blue. This is a trivial task with one blue square, can readily be done with two or three, so the exhibit will explain, and is already quite difficult with four or five, as you will see if you try to do so yourself. Yet, in a world of ever more chosen friends, feeds, and deliberative duties—far more than four or five—people still try to avoid missing something. As Stone has observed: "Information overload. I don't think so. Blaming the information doesn't help us one bit. Information overconsumption. That gets us to the heart of it."[9]

Outcry over divided attention of any sort has generally been escalating. Public safety officials and legislators bewail mounting evidence on the dangers of texting while driving. Educators deplore the distracting effects of background media on gradeschoolers trying to do their homework. The children should be in a room without television, standard assignment sheets stress, with the all too justified implication that, most of time, in the homes of most children, somewhere a television is on. Personal productivity management gurus do a big business telling entrepreneurs not to do everything at once, and to limit multitasking to situations where there are true benefits of concurrency. Neuroscientists explain that, especially for more complex tasks, true multitasking is a myth. Incidentally they also admit that environment has received too little research attention, especially since it has become much more layered in media.

Beyond the Spotlight

Many popular notions of attention exist, as overloaded achievers have discovered through the work practices, meditations, and social networks they have improvised to survive. One reason attention remains such a wondrous but unwieldy topic is that there are so many ways to describe it. Any quick online foraging quickly brings a cascade of views: educators, highly specialized neurophysiology research, Silicon Valley ethnographers, eponymous college courses in psychology, new age personal transformation therapies, or the pioneering opus of William James.

Since attention so often feels divided, perhaps its most widespread popular measure is "attention span": the length of time you can maintain deliberative focus on something in particular. Yet that expression seldom dominates research literature and, even then, means something else anyway. In a world full of interruptions, there is ample mention of involuntary, automatic, or reflexive attention. And there is also attention restoration. Since so many of us wish to protect ourselves against attention theft, there is ample interest in attention control. This coincides with beliefs that attention is something you direct, which in turn coincides with how your gaze moves about. And, for those beliefs, one most popular metaphor is the spotlight.

William James (figure 3.3) famously described attention as focus and periphery. "The faculty of voluntarily bringing back a wandering attention, over and over again," he wrote, "is the very root of judgment, character, and will."[10] Modern experimental psychology has emphasized movements of the eyes (which readily lend themselves to study). The cultural phenomena of things competing to catch your eye make location important, whether on the street or on a screen. The gaze does tend to indicate where attention is. Therefore perhaps the single most widely used adjective for attention is *selective*. In one of his most cited passages, James wrote: "Millions of items of the outward order are present to my senses which never properly enter into my experience. Why? Because they have no *interest* for me. *My experience is what I agree to attend to.* Only those items which I *notice* shape my mind—without selective interest, experience is an utter chaos."[11] This implies that selective focus is voluntary, that it can be maintained, filtered, and owned.

3.3 William James, in the famous photograph by Mrs. Montgomery Sears, ca. 1895 (Houghton Library, Harvard).

But then the eye roves. Wherever vision is focused, it processes the vicinity for where to focus next. At the least-processed level of data gathering, some parts of the retina (called the foveae) are more acutely resolved than others. This makes spatial proximity important. In visual identification tasks, that staple of controllable laboratory-based studies, response proves faster when the target appears closer to the cued center of visual focus. Although the instinct to notice a shiny object in the extreme periphery of the visual field is more often described, the eye continuously processes stimuli close by before moving its center of focus to any one of them, let alone to one more distant.

The wandering eye illustrates the incremental reach so essential to attention. As sociology journalist Maggie Jackson noted in her widely read 2009 book, *Distraction*, "The word [*attention*] is rooted in the Latin words *ad* and *tendere*, meaning to stretch toward, implying effort and intention [*Oxford English Dictionary*]. Even the phrase "attention span" literally means a kind of bridge, a reaching across to widen one's horizons."[12] There is something about switching focus, perhaps especially visual focus, that instinctively satisfies. Something TV producers and advertisers know best of all. Or, in Jackson's words: "The quick cuts and rapid imagery [of television] are designed to keep tugging at our natural inclination to orient toward the shiny, the bright, and the mobile—whatever's eye-catching in our environment. It's ingenious: entertainment that hooks us by appealing to our very instincts for survival."[13]

Extensive research into what catches the eye has made the study of selective attention into the study of reactions to visual events. The mechanisms of vision include zooming in and out like a camera, as well as panning like a spotlight. But, of course, the eye isn't a camera; vision assembles meanings from stimuli by using processes such as production rules, pattern completion, and abstract mental schemas. The workings of attention resemble a series of gateways and building blocks more than they do a spotlight. Although your gaze usually indicates where your deliberative attention has been directed, it tells us nothing about how your attention has been assembled, or its aspects of orientation and habit. To explore these areas, current cognitive research has moved beyond the spotlight metaphor to the role of embodiment in attention.[14]

An Amateur's Neuroscience

You don't have to be a neuroscientist to want a deeper grasp of attention these days. The need for nonspecialists to make their own sense of scientific research on attention these days is only increasing. With the benefit of more time, guidance, and database access than another might summon, this inquiry offers an amateur but hopefully not just anecdotal path through this field. There has to be a way to do so without falling into simplistic universal claims (sometimes known as "neurobabble") how the workings of the brain clearly cause "us" to behave this way or that. There also has to be a way to inquire into attention without just coming away with more overload.

As a way to shape an overview, consider the idea of a mental hierarchy, from senses, to perception, to cognition, to execution, to consciousness. Much of what occurs at the lower levels of this hierarchy never reaches the higher ones. The senses work in parallel, wondrously so, but this parallelism doesn't rise all the way through the hierarchy.[15] Instead, it encounters what earlier psychologists such as Donald Broadbent called "filtering" or what more recent cognitive scientists prefer to call "bottlenecks." More than the lower mechanics of perception or the higher constructs of belief, however, the middle levels of this mental hierarchy appear to interest research specialists most. It is here that sensations are converted into recognition, recognition into action, and action into conscious control, and it is here that cognitive scientists have made considerable advances. They can now model how neurological networks construct thoughts from perceptions, and how representational schemas such as language govern those constructs.

Next, cross the idea of a mental hierarchy with a diversity of purpose. Attention plays many roles. Although cultural concerns about distraction often equate attention with task management, and early psychological research often tested attention through behavior, obviously there are aspects of orientation and intent that neither task management nor behavior addresses. The intrinsic structure of the world may influence these in ways not often researched even now, as if being in the world is less a task than an attitude.

To explain the diversity of attention's purposes, and to understand how these differ in their demands on the mental hierarchy, consider three physiological networks for "alerting, orienting, and executing."[16] These were identified forty years ago by Michael Posner, who has since remained a leader in bringing neuroscience to a broader audience. After interviews with Posner for her book *Distraction*, Maggie Jackson described the three networks metaphorically:

> Orienting is akin to a cognitive mouth, a gateway to our perception, the scout. . . . Alerting is the gatekeeper network, the caretaker who turns the lights on and off and keeps the hearth fires burning in our cerebral house. Simply put, alerting is wakefulness, the cornerstone of sensibility to our surroundings. . . . The executive network is our troubleshooter, a sheriff who moonlights as the judge, with a long reach and a heady power that is, alas, easily corrupted.[17]

As Posner's work has demonstrated, each of these networks has its own pathways in the brain, and much about those pathways develops with experience, not just during childhood but throughout life. Such adaptability invites notions of "digital natives." Yet, even though attention skills do differ among individuals and across generations, as Posner has emphasized, many of these neurophysiological adaptations are culturally and linguistically based.[18]

In any case, the three different attentional networks make different demands on the mental hierarchy. Most likely to consume resources in switching tasks and most subject to direction by effort, the executive network does the least in parallel.[19] Directed executive focus is perhaps the kind of focus meant by "paying attention." It is also the most easily diverted. For example, the executive network often takes the props and cues of one task as reminders of another. At a physiological level, portions of the network's pathways get reused. This can be for tasks as simple as a play on words or as complex as associative reasoning in creative work. Quite often, some nonessential quality such as a name or a color provides the transition point. Wherever an element of the current object of attention also has meaning in another nearby conceptual context, the temptation arises to shift to that one. Plays on words do this, for instance.

Yet those who derive pleasure from keeping watch over as many potentially interesting feeds as possible can mistake vigilance for execution. As research suggests, less can become truly automatic within the executive than within attention's other two physiological networks. It also suggests limits to just how wide a range you can monitor, especially without missing some stimuli.

Some would argue that nothing is salient. There is a famous experiment in which subjects busy counting one thing miss another. Watching a video while two teams, one dressed in white and the other in black, each separately passed basketballs among themselves, subjects were asked to count the number of passes by the team in white. Halfway through the video, someone dressed in a black gorilla suit also walked across the court. This was blatantly obvious, except to those subjects watching the team in white. Busy counting that team's passes, only about half noticed the gorilla suit. Researchers Christopher Chabris and Daniel Simon called this zero-sum effect "inattentional blindness." The "invisible gorilla" became a meme for how no one really knows what attention is.[20]

Almost everyone can recognize the importance of expectations, however. You see what you seek. For evidence that cognitive production relates lower scales of perception to higher scales of intention, experiments emphasize cuing. In the famous "Posner test," response time in recognizing a known shape improves "when [subjects are] first told approximately where it will appear."[21] Similarly, your response to recognizing the appearance of a shape is quicker when you know what shape to expect. You will generally see what you are looking for more quickly than something you aren't. Psychologists call this phenomenon, as generalized among all modes of perception, "priming" or "attentional set."[22] Expectations allocate resources toward a particular kind of filtering.

Resource Allocation

Thus there are very real limits to selective attention. And there are very real advantages to orientation. But if you were to keep just one cognitive science idea in support of better practices of attention, it would be "resource allocation."

What cognitive scientists call "resources" are for the most part brain pathways. At the lower levels of the mental hierarchy, they provide perception. At higher levels, they represent knowledge, which has long surpassed perception as the main topic of cognitive research. Mental models, schemas, goals, and plans extend the notion of attentional set. They apply especially well, and are especially convenient to test, for experiments using work tasks on foreground technologies, such as clicking on a screen. From a generation of such experiments, cognitive neuroscientists now understand much more about the physiological basis of knowledge, whose representations apply adaptive networks of brain pathways in modular recombinations for a variety of purposes. Especially at higher levels, these representations operate on streams of symbols; in effect, there is a language of thought. Indeed, cognition research has, for the most part, grown up by emphasizing how the brain processes symbolic representations.

Where resources apply to attention, their most obvious role is as gatekeeper. (Here one famous metaphor was Aldous Huxley's "doors of perception.") Each level in the attentional hierarchy lets some stimuli through but keeps most others out, all according to rules and receptivities set by still higher levels. Here it helps to imagine what cognitive scientists call a "perceptual store": accumulations of data from the senses. Vision, for example, works from highly fragmented sensory inputs to build identities of objects, spaces, and patterns.[23] To accomplish this may involve buffering, delaying, or comparing different streams of sensory data, as well as invoking higher representations for the pursuit of patterns in these streams. This is where filtering and resource allocation become complementary.

Consider a specific instance of this assembly process. For even the most basic steps beyond the perceptual store, production-rule resources aren't infinite. Thus vision begins with color, extent, and position as quite separate perceptions, and only later, through what neuroscientists call "feature integration," sees shapes among them. In Anne Treisman's famous experiments with reading, the T character took subjects half as long to identify in a field of I and Y characters as it did in a field of I and Z characters, T's horizontal stroke being more easily recognized when it was the only one present. Despite how automatic reading becomes, the T character is still first seen as separate pieces and only later, through resource allocation and following what vision scientists call "production rules," becomes a T.[24]

The coupling of filtering and resource allocation naturally varies from one mode of perception to the next—and from one kind of activity to the next. Take motor activities. More of these can function automatically in parallel than activities of other kinds. You can walk and chew gum, run and throw a ball, or perform any number of other motor activities in pairs at the same time. Yet, even here, some gatekeeping can occur, as when actions compete for some higher systemic resource, such as control of your hands.[25] This makes it hard to keep patting your head while also rubbing your stomach. Likewise, it is hard to make sense of two people speaking to you at once, especially with each speaking in a different ear. If asked what you were just told under such conditions, you are likely to conflate pieces of what the two just said.

Bottlenecks occur where the brain must exchange one cued attention set for another. Switching among different modes, streams, and buffers of perception in itself requires resources. As explained by David Meyer, perhaps the best-known experimenter on switching costs, the brain takes time to change goals, to reload the rules needed for the new task, and to reset perceptual filters. Switching incurs costs all across the hierarchy of attention, but especially at the higher levels, which involve greater use of the executive network. Thus the more complex the tasks, the steeper the switching costs.[26]

Therefore, as Meyer famously asserts, multitasking in any meaningful sense simply doesn't exist. Switching entails lags and inefficiencies, not on the timescale of the tasks themselves (which would halt them), but in fractions of seconds, whose effects are cumulative. Awareness of these deficits is effectively lost, however, in the perceptual and experiential pleasures of changing focus. All that fast switching can feel quite good, and can be mistaken for true multitasking. Meyer summarizes that "multitasking is for fun," not for actually getting things done.[27]

In a widely cited study from Stanford, researchers demonstrated the reality of switching costs. Surprisingly, and contrary to the usual arguments for habituation, "heavy media multitaskers" performed more poorly than other subjects on a set of test switching tasks. They were found to be "more susceptible to interference from irrelevant environmental stimuli and from irrelevant representations in memory."[28]

This finding runs counter to the digital natives' argument that whatever you do most you will do best. Although the science may not yet be developed enough to show just where, clearly there are limits to cognitive adaptation. The brain cannot be made into just anything; it doesn't just become what it consumes.[29] In one of the wisest recent essays on what is commonly referred to as "neuroplasticity," linguist and cognitive scientist Steven Pinker wrote: "Yes, every time we learn a fact or skill, the wiring of the brain changes; it's not as if the information is stored in the pancreas. But the existence of neural plasticity doesn't mean the brain is a blob of clay pounded into shape by experience."[30]

Instead, neuroplasticity means that habits and tools of engagement matter more than was previously understood. Anyone with a studio practice might understand how context suggests intent. This is why yoga teachers advise setting aside a particular corner of the house for meditation. This is why it is hard to cook in somebody else's kitchen. Researchers have shown that even a simple habit like reading the news can be disrupted by a removal from usual circumstance; when you move to another city, your routines may settle in differently.[31] Acquiring habits includes links with context, not only with objects used but also with incidental qualities of sound and light where you stand or sit. The brain constructs correlations through metaphors in the course of direct physical experience and, at a less linguistic level, through links with the properties and usable relationships of things in habitual, stable physical contexts. Contrary to the digital natives' argument, disembodied digital media provide too few such properties to construct useful correlations. And because sensibility alters experience, mindless overconsumption may, if anything, do more damage than good. This is not to say that technology eventually makes you stupid or smart, although it may amplify some of your temporary mental states. Hence the neuroplasticity movement emphasizes perpetual learning through mindful, habitual, skillful, situated practices. Recent findings in neuroscience tend to support the old adage among artisans that what you are really working on is yourself.

Knowledge Representations

One reason why attention has fascinated philosophers, psychologists, and cognitive scientists is because it depends so much on knowledge. Not all the workings of attention rise to the level of conscious thought, however. Which ones do so seems to vary; this makes attention seem more selective than it actually is. Much of the knowledge used in attention is unconsciously learned, unnamed, and unintended. More remarkably, much of the knowledge used in attention does not take the form of symbolic processing structures, which seem to be the core topic of cognitive science. Indeed, tacit, action-based, and externalized forms of knowledge contribute more to attention than they do to most other functions of the mind.

A great deal of knowledge is inarticulable, especially when in use. In music, sports, or many another expertise, you can do things you cannot explain. Tacit knowledge also occurs implicitly in organizations and in social relationships: at a workplace, a cocktail party, or the corner office, for example.[32] In a classic argument against the mid-twentieth-century positivist impetus to proceduralize everything, philosopher-chemist Michael Polanyi showed how tacit knowledge exists not only in music or sports but also in scientific research and organizational memory.[33] Thus business and organizational studies emphasize the value of tacit knowledge held by communities of practice. They find that experts don't so much follow rules as they play situations—the contingent configuration of tools, props, co-present players, and built-in cues and protocols for operation—as a way to bring their environment to attention. For an instance of props, having a prototype on the table shapes what its designers are likely to say. The physicist Richard Feynman once explained that his notes were not a "record" of his thought but one of its components: the work occurred on paper and not just in his head.[34]

Recent findings in neuroscience suggest that tools, props, and settings contribute far more actively to far more tacit forms of knowledge than mainstream cognitive science once understood them to do. These findings bring research back into alignment with earlier, precomputational interpretations of attention. For example, the early nineteenth-century psychologists' affinity for images now makes more sense. In much the same way as language constructs, highly resonant images may serve as anchors and building blocks for knowledge states. Art critic Barbara Maria Stafford called this "form as figuring it out" in her 2007 study of "the cognitive work of images."[35] Form assists the speculative imagination, of how different things might be. Art helps assimilate abundant novel stimuli. The resonance of certain images helps explain some "interpretive blanks in the picture of how the mind works."[36] Neither language alone nor any form of symbolic processing completely explains how mental models develop. Because affect and perception interrelate, works that move the emotions also help coalesce mental models. And by means of such phenomena, many neurological processes operate before language does or without rising to the level of deliberation.

This new stage in the cognitive disciplines strongly affirms activity-based approaches to interaction design. Those justify continued interest in the configurations of sites of habitual activity. This raises questions about the role of embodied information in the ambient. In light of researchers' renewed appreciation of context in the workings of attention, this inquiry next turns from attention itself to its embodiment in physical contexts, and then to the role of built contexts. Here the insights of early, precomputational psychologists seem well worth recalling. William James, John Dewey, and Lev Vygotsky all were quick to point out that people often act before noticing that they are doing so, and without consciousness of rules or names, as if nothing shapes attention more than the habits and situations of mastery. Or, as James famously put it: "The marksman sees the bird, and, before he knows it, he has aimed and shot."[37]

3.	ATTENTION
Main idea:	Many kinds of attention; not all foreground deliberative focus
Counterargument:	Selective attention
Key terms:	Resource allocation, adaptation
What has changed:	Cognitive science of attention
Catalyst:	Costs of multitasking
Related field:	Cognitive neuroscience
Open debate:	Digital natives?

Embodiment 4

To thinkers of any era, at least in the West, the world has seemed manifest; it has assumed its form. How does that condition shape information and attention? It is through embodiment that many themes of this inquiry arise.[1]

What, for example, is the role of embodiment in tacit knowledge? How does tacit knowledge provide a sense of flow, in which attention becomes effortless? How do embodied skills depend on the features of their settings? And how does engaged action use features of context as building blocks for knowledge representations? Such questions turn on a paradigm shift in cognitive science, which is generally known as *embodied cognition*. Its basic idea seems simple enough. Philosopher and cognitive scientist Andy Clark and fellow philosopher David Chalmers call it "the extended mind" or, perhaps more usefully for present purposes, "extended cognition."[2] The word "extended" indicates engaged action making use of external circumstances, which cognition uses much like knowledge representations. Activity theorists have understood this since the mid-twentieth century; cognitive scientists began to confirm it in the 1990s. As Clark summarized then, engaged action occurs in "memory as pattern re-creation instead of data retrieval; problem solving as pattern completion and transformation; the environment as an active resource, and not just a domain problem; and the body as part of the computational loop, and not just an input device."[3]

You aren't your brain.[4] Although some have called cognitive science's shift in research focus from the disembodied mind to active embodied use of external contexts (figure 4.1) a "revolution," if you always found purely computational models of mind somewhat suspect, you might sooner call it a "correction."

Engaged Action

A first look at the workings of attention quickly revealed the importance of knowledge representations in resource allocation. Neither those representations nor the allocation can always be conscious and willful. Attention is not just what you choose to attend to.

To resume the inquiry where it left off on tacit knowledge, turn now to the question of engagement. In a philosophical approach to engaged action generally known as "activity theory," there is a rich legacy of work on tacit knowledge. Many an interaction designer has considered this. Since the 1980s, activity theory has helped to move computing beyond the command line to the graphical user interface and thus into the simple work tasks of millions of nonspecialists. It has also helped to move interaction design beyond isolated explicit tasks to a richer relation of subject, objects, intentions, and contexts and, in doing so, to advance interfaces beyond the desktop into sites and

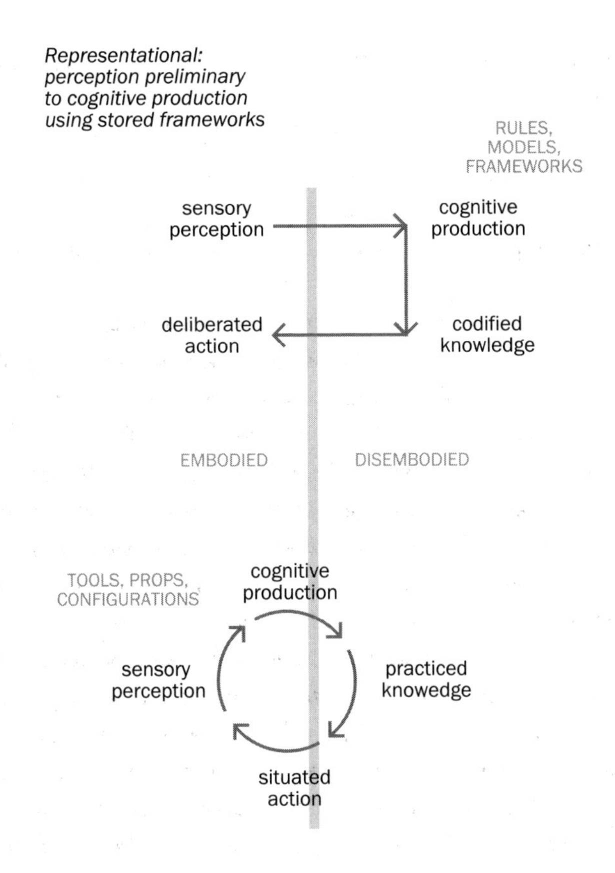

4.1 From representational to embodied models of mind: building blocks in cognition.

situations of everyday life. Anthropologist Bonnie Nardi, among the first to bring work on context and cognition to a wider audience of interface designers, has advocated a paradigm shift from "users and systems" to "subjects in the social world." This has become vital to a new era of interaction design where, unlike earlier desktop graphical interfaces, it is often ambiguous who is a "user."[5]

At its core, activity theory describes a continual rebalancing of internal and external knowledge representations. During what activity theorists call "internalization," learning first makes use of physical props and social cues, as when you learn a skill by watching someone.[6] Conversely, during what they call "externalization," acquired internal processes sometimes resort to tools when applied under new circumstances, such as when you apply them on a different scale or when you get out your calculator. Central to activity theory, these external tools, props, and circumstances act as internal tools. In an influential review of activity theory, Nardi and fellow anthropologist Victor Kapetelinin have explained: "Psychological tools turn mental processes into instrumental acts."[7] The interplay of internalization and externalization allows developed ability to adapt, as to changes of circumstances. This lets experts play situations in an adaptive back-and-forth that gives external components an active role in knowledge, and that continually unites perception, action, and knowledge.

To the skilled tool-using mind, a set of external circumstances becomes "about" something.[8] A floor may invite dancing, just as a rake may invite gardening. As people learn from their settings, they come to associate them with particular states of intent. Activity theory emphasizes intentionality far more than do some competing approaches to the role of external components in knowledge.[9] Intent shapes perception and, with it, discovery of affordances—possibilities for action afforded by objects or environments; conversely, intent is itself shaped by the presence of known affordances.

Because a tool focuses attention on skilled guidance of a specific action, a software object or application that doesn't involve skilled guidance should not be called "a tool." Because a medium gives expression (or bias) to work, it should not be dismissed as "just a tool." Skilled guidance, usually by means of acquired sensory-motor skills, usually puts a tool into relationship with a medium, enough so that they are sometimes understood as one, like painting. Skills and their respective tools are acquired, applied, and shared within specialized settings and communities of practice. A good set of tools has a stable set of features and doesn't benefit from frequent updates. Individual tools suggest their respective uses and may require not so much instruction in what they are for as practice in how well to use them. The craft of using them involves discovering not only finer skills but also helpful configurations and combinations. Props prompt ways of thinking. Surfaces and channels position guidance to tools. Materials invite processes. Again, these may be attributes of traditional form giving but can also pertain to other mediated applications of tacit knowledge. Of particular interest, practice with circumstances builds tacit knowledge. Neural representations of tools and props serve as building blocks for adaptive new uses. Discoveries tend to be incremental: they involve finding not so much one optimal way to do something as a hundred different ways, each somehow appropriate to its physical and intentional circumstances.

4.2 Unmediated affordance, perceived through embodiment: the Warren experiments in stair climbing (1984). Diagram after William Warren.

Casual discovery of useful circumstances depends on a cultivated situational awareness that is different from instructed use of designed and declared means. This is currently a matter of contention among psychologists, or among technology-builders. The influential visual cognition theorist J. J. Gibson held that affordances exist whether designed or not, and are perceived directly, that is, without mediation, in relationship to intent. This was illustrated in William Warren's noteworthy experiment on how subjects recognized that a stack of boxes would serve as a stair (figure 4.2).[10] This pertains to a world where technology too often introduces false (apparent but not actual) possibilities for action, or where procedures sometimes preempt discovery of undesigned, undocumented intrinsic possibilities for action. Easier engagement brings attention to more kinds of intrinsic information.

Habituation

In part, a sense of overload comes from a disregard for the role of habitual context, especially in mastery of action. Any technological activities can become automatic and multitasked, technology advocates argue, just as any dull routines can be improved by introducing media to make them changeable, or if not, at least to tune them out. Yet even to someone who has grown up using multiple electronic media all day, something is lost when nothing persists.

The more enduring the environment, the more it shapes expectations without saturating attention. After regular exposure, unneeded stimuli become less distracting. With regular use, skillful actions become more subtle and efficient. Technically, this involves encoding. The brain builds constructs to allocate resources for frequently encountered processes so that they no longer require so much higher-level deliberation.[11] When orchestrated across a frequently associated set of operations, such built resource allocation helps an activity become automatic; it can become modular when some of these resource "chunks," or pattern encodings, prove useful for other, unfamiliar operations. This suggests why (albeit at some higher level) one practiced skillful activity leads to another, and to greater reflection.

Many of the most natural activities consist of mastering and configuring habitual contexts. As in a good kitchen, the arrangements of utensils, materials, and processes reduce the need for explicit models of what must happen. A recipe may provide a point of departure, but what distinguishes expert cooks from merely competent ones is their readiness to depart from it. Much of this is habituation to context; it is usually harder to cook in someone else's kitchen.

Habit need not be dulling. Habituation allows more perceptive engagement of circumstantial differences in the play of expertise, from day to day, site to site, or group to group. Contrary to the assumptions of a mechanistic industrial age, a great deal of social and organizational knowledge gets embodied in routine.[12] Or, as William James so aptly put it: "Habit is thus the enormous fly-wheel of society, its most precious conservative agent."[13]

Habit expands the context against which newly changed stimuli may become salient. Richer, more tangible media can play a role in this. By contrast, disembodiment can lead to deskilling and overload. Overmediation and continual flux may preempt deeper, less mentally taxing forms of engagement.

Runaway Ontology

In philosophy, ontology generally studies existence, especially categories of being; but in cognitive science and software design, ontology is the process of specifying categorical schema, such as a hierarchy of software classes, in which modularity and reuse are particularly valuable, more so than a simpler vocabulary or taxonomy. Indeed, ontology has become so ingrained in information technology through the object-oriented software design that today's version of the categorical imperative involves declaring tags, variables, and methods for just about everything. You might call this "runaway ontology," though few others have thought to do so.[14]

Although cognition researchers find ontology very useful for understanding the modularity and semantics of knowledge representations, they increasingly recognize that attention doesn't always rise to the level of formal symbolic identities (figure 4.3).[15] Activity theorists sometimes call this awareness "preontological." Whereas computer scientists focus on categorizations, object hierarchies, and nomenclatures in their work on knowledge representations, activity theorists have shown how many practices occur without such explicit semantic schemes.

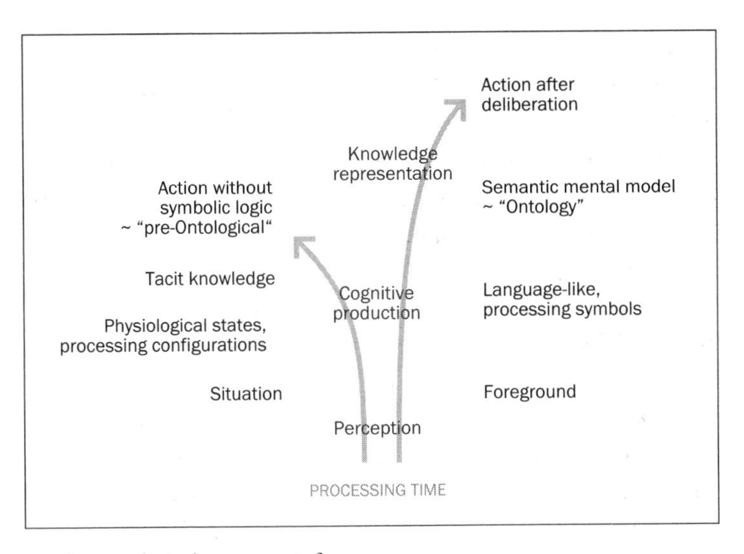

4.3 Pre-ontological awareness is faster.

Like any other organ, the brain operates biologically, by means not unique to humans. Although Aristotle explained how the body has lower mechanisms tying attention to emotion, recent works in embodied cognition have debated just how. For example, Anthony Damasio's influential 1994 book, *Descartes' Error*, argued that the brain and its deliberations are an addition to, not a replacement for, animal biological attention.[16] Melvin Goodale and David Milner's explanation of "dorsal pathways"[17] had previously argued that tacit environmental perception employs different processes, even different areas of the brain. Thus one patient who had lost the ability to recognize a ball on a table could nevertheless catch it when it was thrown. Much as foreground vision has a seam where the two eyes' parallax produces anomalies, so background vision has a seam at arm's length. Embodiment is essential to nondeliberative perception and thus to tacit knowledge as well. Such perception happens more quickly than reasoning. This explains why, like William James's marksman, you can act before you become consciously aware of doing so.

From the experiential perspective of engaged activity, runaway ontology could be counterproductive. Assigning names, instructions, and procedures to everything could prevent more intrinsic circumstances from contributing to the play of tacit knowledge.[18] Without stable tools, shared sites and configurations, and the capacity to move between internal and external representations, activity may not feel engaged. When external representations are confined to disembodied digital media, which require deliberative focal attention, or when their designs assume that no one is paying attention or has the basic skills to engage undeclared affordances, a more tacit awareness based on embodied engagement may not develop.

Situational Awareness

Soldiers and surgeons alike tend to prize their "situational awareness."[19] The ability to read a context on the fly can mean life or death to these experts. Where a merely competent person would play by the rules, an expert knows when not to and directly plays the situation. Such agile awareness of course also counts in many smaller, less critical situations of life. Many disciplines whether in computation, cognition, linguistics, organizational studies, or interface design, now emphasize these as an alternative to more formalized models of knowledge.

Numerous, less critical applications of situational awareness occur in industrial robotics. Reporting from the experience of that field in the 1990s, activity theorist Andy Clark explained the disadvantages of deterministic external models relative to the then-new approach of adaptive neural network learning. In doing so, he renewed an age-old goal of cybernetics, namely, to build perceptive technological control systems. Central to Clark's contribution, engagement with context provides an active resource, and not just a starting input, for processes of movement, memory, and other applications of executive attention. On the role of structure, Clark observed: "In general, evolved creatures will neither store nor process information in costly ways when they can use the structure of the environment and their operations upon it as a convenient stand-in for the information-processing operations concerned."[20]

To use the environment as an active resource means that skills can neither be acquired nor applied nor explained without it. This is a fundamental premise of tacit knowledge. It is also a fundamental principle in developmental learning, one that has been around cognitive psychology for fifty years, in the work of Lev Vygotsky, most often considered the father of activity theory.[21] The tool metaphors in software interfaces owe much to his work. The later consensus that structures of language indeed shape pathways of cognition makes Vygotsky seem prescient. But it is the importance that activity theory attaches to structures of physical tools and situated tasks that interests us here. For quite some time, it has been thought that a tool, or even its setting, such as a digital craft studio, is not only for some purpose but also about it.[22] With respect to the capacity of intrinsic structure to influence attention, this means that a set of tools might both put someone in a frame of mind and be essential for building that frame of mind in the first place. Clark referred to this learning effect as "scaffolding."[23]

Much as a physical scaffolding allows a more permanent structure to be built or maintained, so a mental scaffolding allows a situated action to be developed and performed. In a critique of robotics, Clark summarized the basis for what are now more widely used approaches to emergent patterning (computational neural network learning) of actions, especially complex coordinated motion: "memory as pattern re-creation instead of data retrieval; problem solving as pattern completion and transformation; the environment as an active resource, and not just a domain problem; and the body as part of the computational loop, and not just an input device."[24] This resonates well across phenomenology, which understands the world by what can be done with it, cybernetics, which seeks a biological metaphor for sensate technology, and digital craft, which finds a consummation (i.e., pattern completion) in the flow of engaging a medium.

More recent complexity theory has shown how a dog making its rounds needs only to connect with a few neighboring dogs using just a few messages, all of them in the physical setting that they share, for remarkable actions to emerge. Not to engage surroundings would impair such a system; that would be unnatural.

We have adopted the expression "intrinsic information." More generally, *intrinsic* means "within the essential nature of something." Thus passage is intrinsic to a door. Intrinsic properties persist as long as the things to which they belong do. Physical persistence provides mental building blocks. Without persistent structure, the world seems less knowable. Without the recurrent, familiar structure of phenomena as a basis, many pattern-based aspects of cognition have less to work from. Intrinsic structure has a place in epistemology. And nothing provides intrinsic structure quite so well as built environments.

Embodiment provides orientation. The setting in which you are present has the marks and sometimes the presence of others. Especially when you know a setting through habitual use, it tends to cue your intentions, memory, and perception.[25] The very configuration of people, places, and things has significance. The persistence of some such configurations holds meaning. Thus there exist cities of collective memory, halls of assembly, and hallowed commons. Not only architects and planners know this. Each person, every group, and any institution has some story. The city is made of stories, which you recall by walking. The city is also made of signs of these stories, whether inscribed or implied. A particular doorway or the stream of pedestrians passing through it might recall an event or represent a wish.

Architecture, as an aesthetic production, can raise situational awareness to the level of atmosphere, and thus make it part of the ambient. In a 1993 essay on atmospheric aesthetics, now cited across a wide range of disciplines, philosopher Gernot Böhme explained:

> The new aesthetics is first of all what its name states, namely, a general theory of perception. The concept of perception is liberated from its reduction to information processing, provision of data or (re)cognition of a situation. Perception includes the affective impact of the observed, the "reality of images," corporeality. Perception is basically the manner in which one is bodily present for something or someone or

> one's bodily state in an environment. The primary "object" of perception is atmospheres. What is first and immediately perceived is neither sensations nor shapes or objects or their constellations, as Gestalt psychology thought, but atmospheres, against whose background the analytic regard distinguishes such things as objects, forms, colours, etc.[26]

A sensibility to architecture, its situations, and its atmospheres gives much else continuing relevance in an otherwise low-resolution world. The appreciation goes beyond the appeal of form, and is as much about persistent inhabitation. A reflective awareness that shapes spatial configurations of everyday practices may be valued in itself: as a way of engaging intrinsic information, as a community of practice, or as a way of calming.

Effortless Attention

For understanding the craft-like experience of engaged action, Mihaly Csikszentmihalyi's oft-cited theory of "flow" explains a clear basis for calming. It also provides a clear counterpart to the focal, deliberative, symbolic processing kinds of cognition that preoccupy most cognitive science research. In a process that anyone can recognize, flow occurs when you confront a challenge that you can mostly master with skills that you mostly have. Flow involves knowing how rather than what. Thus its play is tacit—you may not be able to say what you are doing. Most important, in flow, you can meet a higher demand without feeling an increase of effort.[27]

"Effortless attention" deserves much more respect.[28] It is too often mistaken for purely automatic attention, which is different and has often been the goal of so much behavioral engineering, as in conditioned response to repeated stimuli. Behavioralism has given context-and-cognition research a bad name, for it reduces individual agency and ignores the importance of intent. Automatic attention has also become an easy excuse for abuses of multitasking, as if any act becomes automatic if you do it often enough. As should be evident from our look at resource switching and attentional set, that isn't the case. Few acquired tasks become fully automatic. Instead, they develop a subtle sensitivity to context, in which attention is increased and purposeful, but effortless.[29]

To the philosopher Brian Bruya, who has interpreted Csikszentmihalyi as a part of a larger East-West dialogue on effortless attention, sensitivity may be the key. "Rather than a spotlight, or a filter, and so on, this model posits that attention may be profitably conceived of as a mechanism of sensitization that draws information relevant to dynamic contextual structures of reference through dynamic processing pathways."[30] The model depends on purposeful engagement with persistent context: "Occasionally, activity domains stabilize as temporary predominant structures, inhibiting competing structures of reference by virtue of the activity's autotelicity, thereby allowing for sustained, focused attention that feels effortless."[31]

Perhaps because of heightened sensitivities, effortless attention can be rewarding, even restorative. Deliberative focal attention (the executive network) gets overworked, especially when interruptions become more numerous and less predictable, and when other kinds of attention diminish. Sitting in the dark with a screen is a perfect recipe for this problem. Going for a walk in the park, or at least sitting with a view of some trees, seems like the easiest cure. Attention restoration theory was introduced by the psychologist Stephen Kaplan and the landscape architect Rachel Kaplan.[32] This theory emphasizes fascination, especially with nature. Fascination can be involuntary, especially when "soft" (as in outdoor recreation) rather than "hard" (as in a hobby.) This the Kaplans contrasted with voluntary "directed" attention, which suffers from fatigue. Attention restoration thus becomes an important agenda for those whose cognitive apparatus has somehow been damaged, such as by years of dealing with poorly designed technology.[33] Environments where more engagement occurs with less effort can help with better practices of attention. It seems fair to say that attention restoration, whether in formalized research or weekend sports, taps some latent abilities and preferences.

To what extent can artifice be restorative? Why else are there so many movies and games? After a day of drowning in unpleasant data being dumped on them by others, some people relax with an evening of drowning in more pleasant data of their own choosing. But when it comes to restoring the powers of attention, might going for a walk do more good? Can a walk in the city do as much good as a walk in the woods? Or is looking at high-resolution images of the woods on a plasma display good enough?[34]

Embodiment makes the difference. Walking provides more embodiment, more opportunity for effortless fascination, and better engagement than looking or sitting. Depending on the balance of fascinating and annoying stimuli, a walk around town may well do some good. That balance is now in play, under the rise of the ambient. This topic will resurface throughout this inquiry.

For now, with respect to the workings of attention, simply note the distinction between attention as something you pay and attention as something that flows, a distinction subtler than the distinction between voluntary and involuntary. Intentionality, sensitivity, conditioning, and contextual cues usually enter the process of attention. Engaged action occurs without the use of explicit knowledge representations, and with a diminished role of the executive among attention's three networks.

In particular, note how effortless attention becomes an end in itself. To describe this, Csikszentmihalyi popularized the word *autotelic*. As Bruya explains, in autotelic experience, "the activity provides the impetus for action, involving a challenging activity that requires skill, the merging of action and awareness, clear goals and immediate feedback, concentration on the task at hand, a feeling of being in control, a loss of self-consciousness, and an altered sense of time."[35] Losing yourself in the work is a gain and not a reduction, even at the neurological level. The psychology journalist Winifred Gallagher has investigated this gain through the many indirect benefits of "choosing the focused life." In the interplay of voluntary and involuntary attention, it is possible to cultivate bias, as in examples Gallagher used, to noticing birds. Selective attention cues particular sensibilities, which then cause involuntary attention to pick up on some stimuli, and not on others. This effect is cumulative, and leads to fascination. Plants fascinate gardeners; building details fascinate architects; emergency rescue vehicles fascinate young children. In this regard, paying attention simply means keeping a commitment to stay fascinated about something. It is the opposite of being blasé, that jaded state of overstimulation held to be the default state of modern urban subjectivity. As a cancer survivor,

Gallagher learned firsthand how choosing to be fascinated affects health and motivation. Top-down choice of sensibility affects which kinds of bottom-up stimuli become salient, which reduces the drift and capture of attention by whatever comes along. Thus, being fascinated by their respective pursuits, birders, gardeners, musicians, and architects alike have less use for the latest media event, the newest gadget, or a background television feed. Effortless attention may make furnished entertainment irrelevant, but we can only hope that a surplus of furnished entertainment won't make effortless attention irrelevant.

Although there are neuroscience debates on this, many fields find it obvious enough: cumulative, cultivated engagement with embodied, direct stimuli provides a defense against predations of mediated novelties and interruptions. As Gallagher summarized: "Like fingers pointing to the moon, other diverse disciplines from anthropology to education, behavioral economics to family counseling similarly suggest that skillful management of attention is the sine qua non of the good life and the key to improving virtually every aspect of your experience, from mood to productivity to relationships."[36] Yet this isn't just positive thinking, and it isn't just growing a thicker skin. For across these disciplines, and in meditation practices as well, filtering isn't so much a tuning out as it is a tuning in. Commitment to a bottom-up sensibility makes many other things more vivid, and mitigates overload from too many top-down demands.[37] In particular, it cultivates bias toward intrinsic information as a basis for flow. Effortless attention occurs amid practiced engagement with a medium, whether the soil, a musical instrument, or your favorite design software. It becomes craft. To live well is to work well. Engaged, skillful experience makes better citizens.

Taking Form

To summarize where this inquiry has gone so far: Ideas of the ambient began from a philosophical basis in embodiment. Tangible, embedded, and ambient interfaces have now become usual, and appear at street level. Periphery, which you are aware of through embodiment, has become much more important in the use of information technology. Information, though normally assumed to be disembodied, especially when defined as signals sent, is more accurately true semantic content, which often depends on and refers to a situation. At the same time, the intrinsic structure of situation may itself inform, without anything being sent. Thus the workings of attention include not only foreground deliberative processing of symbols and messages, but also peripheral orientation to prospective stimuli, and not only as inputs but also as building blocks to cognition. Through embodiment, attention engages undeclared affordances that engage nameless actions. Embodiment makes it easier to build tacit knowledge, especially through habituation. Physical elements often anchor and resonate across abstract processes of cognition. Under the right circumstances, increases of attention come with perceived increases of effort or distraction.

Recall how *embodied information* is at least twofold. First, being in the world, it informs through the intrinsic structure of situations, that is, without mediation. And, second, all the mediation that is usually meant by "information" increasingly appears in the embodied contexts of everyday life, where it assumes the form, or becomes a feature, of familiar objects such as cars and parking meters. This creates new prospects for form that informs. In a project by Realities United (currently on hold), the ancient medium of smoke signals is reapplied to indicate the frequency of

4.4 Embodied information: emissions indicator in the form of smoke rings for a proposed waste-to-energy power plant (architect Bjarke Ingels/BIG). Photos courtesy of realities:united.

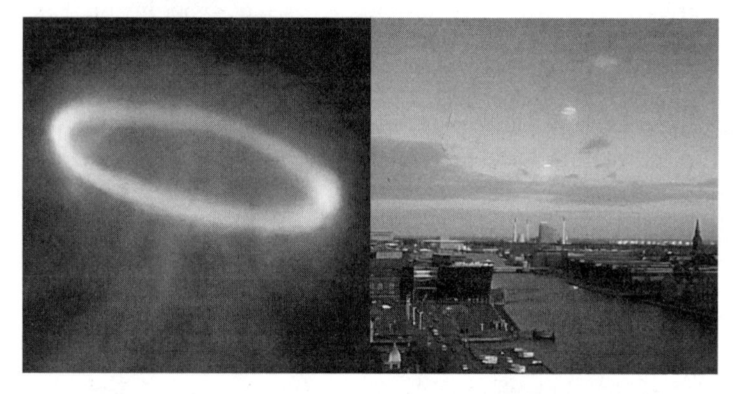

Smoke rings, Mt. Etna, Italy. Photo: ©2000 J. Alen.

Procession of smoke rings seen from Børsgade, Copenhagen. Design and montage: realities:united.

CO2 emissions from a waste-to-energy plant (figure 4.4). This design has reduced cognitive load by assuming the form of a physical object. This is not just data visualization but also data formation. It is an interface you do not operate, and as a part of the scene it is ambient. Such developments in ambient interface may as yet be a sideshow in comparison to how disembodied information media blanket urban space with their screens, but it's a start. As yet, the capacity to tag, to project, or even to inhabit one's own contributions or one's group's curations of augmented urban space is at a very early stage. The challenge is to find the right contexts, scale, texture, timescale, and spatial resolution, and then, as this inquiry attempts, to combine insights on attention with insights on the history of the built environment. For all of this prospect, it seems wise to note that information can take form.

4.	EMBODIMENT
Main idea:	Embodied cognition
Counterargument:	Representational theories of mind
Key terms:	Pre-ontological, runaway ontology
What has changed:	Cognitive science of engagement
Catalyst:	Deskilling and distraction
Related field:	Activity theory
Open debate:	Tacit knowledge?

Fixity 5

What is it like to approach the topic of attention from the perspective of architecture and urbanism? How do the fixed forms of architecture and the city channel the flux of ambient information? Consider this role in the more general sense of the word *architecture*. Beyond its specific meanings for the design of buildings, this can mean any deliberate, irreversible configuration that organizes and confers identity on its inhabitants. So you could say that architecture's cognitive role begins with how it fixes activities, even communications, into persistent arrangements (figure 5.1).

In yet another instance of superabundance, countless historians, theorists, and critics have contributed interpretations on the relationship of media, architecture, and the city. Many have explored the role of architecture in the media: as film sets, television backdrops, spaces to explore in video games. Conversely, many others have explored the role of media in architecture: radio, then television in the home, signage on building facades,

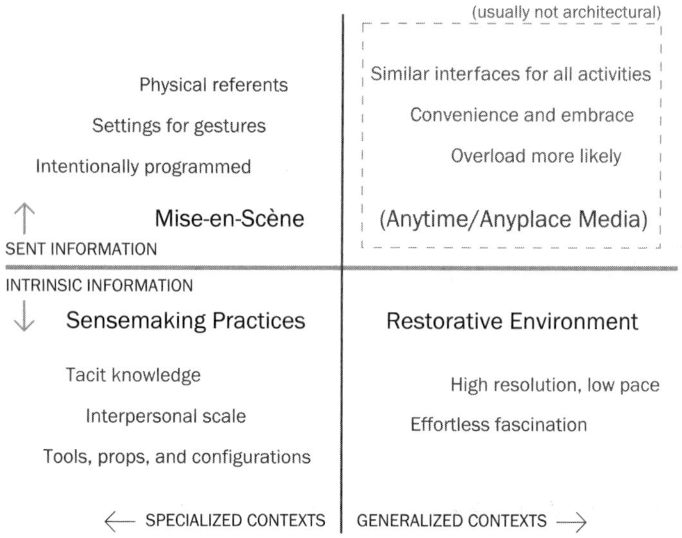

(usually not architectural)

SENT INFORMATION ↑	
Physical referents Settings for gestures Intentionally programmed **Mise-en-Scène**	Similar interfaces for all activities Convenience and embrace Overload more likely **(Anytime/Anyplace Media)**
Sensemaking Practices Tacit knowledge Interpersonal scale Tools, props, and configurations	**Restorative Environment** High resolution, low pace Effortless fascination
← SPECIALIZED CONTEXTS	GENERALIZED CONTEXTS →

INTRINSIC INFORMATION ↓

5.1 Architecture's cognitive roles.

smart green buildings. Most of these note how successive layers of communications media transform the role of building, but never fully replace it.

To look ahead in an age of the ambient, such studies of media in architecture now invite environmental histories of information. This inquiry attempts a start at one, in the book's second half, to complement this half's survey of attention. Before turning to that, however, first consider architecture from the perspective of attention, especially embodied cognition. Fixity provides a distinct way into that. Although usually taken for granted and sometimes sought to be overcome by fashionable kinematic installations, fixity distinguishes architecture. Few other works of design are intended to hold still, and to last so long. Once this was valued more; many scholars and cultures have equated civilization with permanence. Yet even today when almost nothing seems permanent, the persistence of some forms relative to others still matters. An age of flux still needs host forms and channels, much as a river needs banks. Instead of "permanence," which by now is a questionable word, it is better to describe this configuring quality as "fixity."

Mise-en-Scène

For a well-institutionalized example, consider a courtroom, where almost everything has been prescribed. The courtroom demonstrates the principle of "speaking in place."[1] Where you sit affects what you may say or what may be said to you. Expectations are altogether different depending on whether you find yourself in the witness stand, the public gallery, or the jury box. Where things are written down in the courtroom also creates lasting consequences. External feeds such as video cameras may be prohibited. Inscriptions carved into the cornices may exhort citizens to put aside their own interests and to serve a higher common good.

It is no accident that the courthouse is the grandest structure in county seats across the United States. These served as the main units of political organization on the frontier, and their courthouses were often the first sign that settlers had come to stay. Today, when Victorian courthouses are torn down in favor of generic glass-and-metal buildings with more convenient parking, something surely is lost. Civic identity has to count high among architecture's roles. Although a distinct civic structure also has more overtly symbolic functions—you can tell a courthouse from a church at a glance—it operates best by tacitly shaping public proceedings. It is also no accident that Winston Churchill's ever-cited remark on the social bias of buildings, how "thereafter they shape us" was prompted by a proposed reconfiguration of a floor for formal public speaking, the war-damaged chambers of the House of Commons.

More mundane architectural venues such as the bistro and the conference hotel likewise come with varieties of configurations for sitting, listening, chatting, or grandstanding (as with karaoke or slides). These, too, take cues from the size and shape of tables and rooms. They demonstrate how even in an age of text messages, social relations still rely on interpersonal distance to affirm and constantly reinterpret themselves. Power brokers and socialites alike know how to work a room, and what it means to stand or sit a few inches closer to or farther away from one another.

Specialized designs not only arrange people for particular purposes, but also provide them with useful configurations of props and tools. A hotel room comes with expected amenities. A retail shop floor draws in customers with a carefully studied sequence of tables, racks, and display cases. A design center might couple sleek front offices with a backroom mix of toy-strewn personal stations, casual meeting spaces, material resource bins, and prototype assembly zones. A hospital presents enough complex architectural requirements, whether for adjacencies, technology support, or access control, that, as a specialization, hospital design keeps any number of architecture practices busy.

Conduct often depends on information fixtures as well—not just places to plug in and screens to project on, but also barcodes to scan, cards to swipe, and, even on the simplest of objects, cautionary instructions to accept. As these markup languages proliferate, it is worth understanding the implied significance of their placement. How do fixed inscriptions complement mobile communications? When does inscription reinforce more intrinsic information in the configuration of a space, and when does it just get in the way?

Architectural computing pioneer William Mitchell once explained this design challenge as a matter of "placing words." To shout "Fire!" in a crowded theater means one thing, he began; to shout "Fire!" to a squadron of soldiers means another, to affix a "Fire" label to a plumbing outlet is to tell firefighters where to hook up a hose; but "if I receive the text message fire on my mobile phone, at some random moment, I can only respond with a puzzled huh?"[2] The point is that many words refer to, or take scope from, where they are exchanged. A dramatist might understand this as "mise-en-scène":[3] a script needs a setting; objects (props and backdrops) provide orientation. A neuroscientist would understand this as "scaffolding": the objects and settings that language refers to become building blocks of action and thought. Rich metaphor depends on analogy in spatial experience. Indeed, imaginative thought most often works in metaphors. Such mise-en-scène was simple enough before telecommunications. "The meaning of a local, spoken, synchronous message is a joint product of the words, the body language of the participants in the exchange, and the setting." Mitchell explained.

> But the introduction of technologies for inscribing physical objects with text, and the associated practices of writing, distribution, and reading, created a new sort of urban information overlay. Literary theorists sometimes speak of text as if it

> were disembodied, but of course it isn't; it always shows up attached to particular physical objects, in particular spatial contexts, and those contexts—like the contexts of speech—furnish essential components of the meaning.[4]

Electronic media complicate this overlay. Ubiquitous technologies exacerbate this anytime-anyplace problem. Music and television feeds show up where they are not welcome. Mobile phones lead to inconsiderate use. Self-service touchscreens with network connections mediate ever more kinds of transactions. As ever more diverse media cut ever deeper into everyday life with links and portals to someplace else, people suspend may not only disbelief about where they are at the moment, but eventually also a more general sensibility to surroundings.

Mark Twain has left one of the earliest surviving essays on the oddity of hearing half of a phone conversation.[5] Early critics of radio protested how it separated performance and reception, and created a preference for self-contained content that could be repeated anywhere.[6] Whereas cinema was conceived with the expectation that viewers would sit down to it, and it alone, in immersive, specialized theaters, programming for radio and television had to be produced with no clear expectation about how it would be received. This was true not only for speech but also for text and images. These new relations of media and the city eventually led to new abuses of power, especially once radio reached everywhere, a reality that totalitarians ruthlessly exploited. Media-enabled power structures recast the role of architecture in the process, or at least forced architects to rethink it with respect to broadcast distractions. By the second half of the twentieth century, such relations became the core focus of social critiques, as by the Frankfurt school in postwar Germany or its postmodern interpreters amid the runaway consumerism of late twentieth-century America. Alas, these critiques have played up architecture's part in spectacle to the neglect of its quieter effects.

By contrast, a more basic reading on context and cognition would emphasize the role of placed words, images, data feeds, and network links as active components of everyday experience. It would emphasize the rising cultural importance and attainments of the discipline of interaction design, where the aesthetics of everyday experience depend less on prepackaged units of transmitted entertainment and more on the discovery, navigation, and engagement of local affordances. After a long hiatus in the one-size-fits-all broadcast age, the relation of media and the city is coming back around to more helpful role as mise-en-scène.

Making Sense through Form

Form informs. The two words obviously relate. Yet this simple truth gets forgotten when information is always considered disembodied. Consider an illustrative case from everyday life. Two elegant doorways set side by side in a carved stonework link a deeply shaded courtyard with a long, tall reading room, finely detailed in inlaid wood. To be in one of these spaces is to feel its connection with the other, to find that other inviting. The design and placement of the doorways are one of those architectural delights which anyone can understand and which attract visitors to a university campus. You would feel like walking though one of the doorways, whether you needed to or not.

Although the courtyard is plainly visible through the leaded-glass panels of the doors, local life-safety authorities have recently added a bit of instruction (figure 5.2): a standard exit light, mounted at eye level on a stand between the two doorways and

5.2 In case you didn't notice . . . : an exit sign, in the age of compliance.

facing into the building. Perhaps it was required by new federal regulations for emergency preparedness or, more likely, by requirements to honor a newer building code when doing other work on the older structure. The sign's effective message, however, is that, unlike in times past, today somebody would not otherwise notice the outdoors through the glass door in front of them. By building in such a connotation, in effect signs like this institutionalize distraction.

Elsewhere in the world, the law still considers some spatial configurations self-evident, as if common sense still exists. For example, the canals of Amsterdam not only lack warning signs about the possibility of falling in; they also lack railings. Yet almost no one falls in.

So the lesson is this: in the rush to ambient information, don't forget, and don't cover over, the natural meaning of things. Built form plays an important role in helping make sense of everyday life. Not to know anything through acquaintance would be a great loss. And, besides, how can you expect others to raise their environmental sensibility and reorient an entire political economy to a changing planet, if you can't expect them to sense, uninstructed, that a door with a clear view of the outside is an exit?

Much as form in-forms, similarly to in-habit habituates (figure 5.3). The more enduring the environment, the more it shapes expectations without saturating attention. To dwell means not only to reside but also to maintain a bias of focus. You might dwell in a space, and dwell on a sensibility. By contrast, without any persistent context, you are nowhere. Relentless superficial flux of media undermines your sensibility. Without habitual practice in intentionally configured settings, you will find it hard to achieve a sense of flow. We have examined habituation with respect to tacit knowledge and seen how important scale is to that. You don't need instructions to know that a stair is for climbing, an arcade is for strolling, or that a tower asserts the prominence of its owners. Something you glimpse from afar differs from something you examine up close. Something you inhabit differs from something you hold.

Thus, according to the cognitive principles presented thus far, environment is not an other, nor an empty container, but a perception of persistent possibilities for action. Fixed configurations of spaces, props, and artifacts support activities in ways

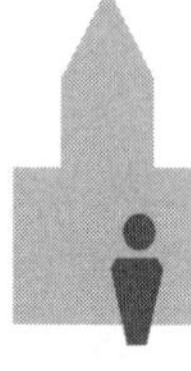

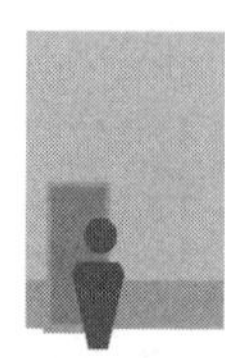

5.3 Ways that form informs.

that go beyond housing them. Unmediated affordances help shape tacit attention. Although habit can assimilate and internalize almost any process, it seldom does so automatically.[7] Habitual action still makes some demands on executive attention. An increase in tacit knowledge involves an increase of attention but a decrease of perceived effort. Where mediations are necessary, they are best made unobtrusive. Better interface design works with the embodiment of tasks and the configurations of the spaces that support and constrain tacit knowledge.[8]

Note how this approach shifts research priorities from behavior to sensibility. Far too often, the goal of research into environment and cognition has been to induce behavior. A great deal of space gets socially engineered, whether for retail sales per square foot, test scores in classrooms, personal safety on sidewalks at night, or production rates on the factory floor. Industrialization left a bias toward procedures and against more tacit, habitual mastery. When, in the mid-twentieth century, the social sciences sought to emulate the predictive certainty of the physical sciences, the discipline of architecture attempted to apply analytical behavioralism, for example to the design of public schools and housing projects. This proved to be inappropriate and unwelcome, however; the disciplines of architecture and cognitive science have largely avoided one another ever since.

Counterarguments to behavorialism tend to emphasize intentionality. Especially over time, to inhabit is to habituate. Inhabitants and spaces mutually adapt. When inhabitants share a perception of affordances, that reinforces an identity of the environment. As inhabitants learn from their settings, they come to associate them with particular states of intent. Drawing on accumulated experience of intent, they develop more abstract mental models, which then become second nature. Here again is the cycle of externalization and internalization that lies at the core of activity theory.[9] With this focus on intentionality you don't have to take up a behavioralist agenda to study cognition.

Organizational theorists use the term *sensemaking* to describe the use of context for coping with complexity and overload. Most of this context is social: mindful workers cue off each other to figure out what is going on, and what they should be doing. But some of it is physical, most obviously because specialized configurations of spaces represent what has worked before, and more subtly because the sensemaking process depends on other kinds of retrospect. Although it may seem like a truism that objects and settings bring forth courses of action, it is also something that interaction designers have studied for decades.[10] The idea of circumstantial retrospect has been acknowledged as an important contribution to the principles of embodied cognition.

Because more habitual contexts make new cues more salient sooner, it is important not to cover them over in too many overt media and procedures. As noted in the previous chapter, runaway ontology can be counterproductive. Policy has its limits, beyond which latent affordances must be given play. In the words of Karl Weick, the most recognized scholar of this process: "As sensemaking unfolds, at least seven resources—social context, identity, retrospect, salient cues, ongoing projects, plausibility, and enactment—affect not only the initial sense one develops of a situation, but more important the extent to which people will update that sense." This updating demonstrates the role of contextual particulars. "Action is always just a tiny bit ahead of cognition," Weick has explained, "meaning that we act our way into belated understanding."[11]

As architecture arranges interpersonal distances in space, configures everyday processes, represents organizations, and shapes everyday habits within them, it also inobtrusively supports sensemaking. It does so with quiet alcoves and grand halls, with private gates and public plazas, and with specialized sequences of rooms. It tacitly cues what to say where, how to act in groups, and toward what goals these arrangements have been institutionalized. Architecture provides scaffolding for knowledge and expertise, and so contributes to better practices of attention.

Persistent High Resolution

Amid escalating experience of flickering low-resolution ephemera, carefully composed duration and rest may seem welcome. This makes architecture's greatest asset its persistent high resolution. Besides furnishing mise-en-scène for communications and scaffolding for sensemaking, fixity provides exercise and rest for attention. It does so by means of high resolution with low demands, through something as simple as a view of sun crossing a wall. As noted with respect to attention restoration research, this helps explain how to avert a sense of overload. Access to natural phenomena may restore attention best, and participation in art or sports may help as well—hence these are called "recreation"—but what of more living with the details of buildings?

Consider a simple stone wall (figure 5.4).[12] The texture of stone appeals to the eye. You could study it at length, or just keep it in background view all day. A seat with a view of this wall would be a better place to work all day at a laptop than a seat with only views of painted wallboard, or of textile-covered steel cubicles. Orientation would matter: the slow change of light through the day, or the slower change of the color of moss over the seasons, would improve the effect. More subtly, immediate orientation would suggest larger orientation within the city and the world. You might know which way is east, and that might somehow come to matter. Meanwhile, you might receive some cultural orientation from posters that come and go on the wall. And you might have a different reaction if somebody signs the wall in spray paint one night: it might feel like a transgression. So might a corporate advertising campaign. If a local billboard company mounted an electronic display on the wall, you might feel angry, helpless, or hurt. But even if you could mount your own display there and, in the limit case, cover the wall with pixels on which to display whatever you like—perhaps Parisian boulevards one day and Montana forest scenery the next—you probably wouldn't. And you probably wouldn't even if the technology were provided at no monetary cost. For the cost in attention management would be high. And, in the long run, the play of shadows on the stone does a better job than any digital wallpaper could at balancing your attention. No pixel technology could

5.4 Persistent high resolution.

deliver quite such high resolution, nor be expected to change its appearance quite slowly enough, nor to have such fixity.

The word *patina* helps capture this sense of slowly changing, irregular, accumulated texture. In the case of stone, weathering alone will provide some of that texture. To sculptors, patina may mean deliberate tarnishing, as of bronze by acid wash. To others, patina may mean a buildup of oils, irregularities in the patterns of regular cleaning, or subtle erosion of material by wear. Makers of blue jeans and designer furniture may carefully "distress" their wares before selling them. Architects often explain how weathering is a feature, not a defect. For, besides exposure to the elements, weathering can reflect patterns of use, as in the gradual wear of stone steps. Such accumulated traces may afford learning, cultural traits, and acquired perceptivity at a more abstract societal level.

Without some persistent structure, attention diminishes. Total novelty disarms. Better to have serendipity amid practice. Thus it makes sense that there is no better basis for cultural practice than the enduring city. You almost can't have one without the other. As you return to it, across your lifetime, a great city reflects how times have changed, and how you have changed. Later in life, these reflections start feeling more important than new sights, although to the seasoned traveler, new sights also reflect better-known ones. You might trace a chosen theme across the world in its subtle differences and its changing relationships to more persistent, recognizable settings.

How does the persistent high resolution of the built environment complement the jumpy low resolution of the screen? How, instead of everything happening anywhere, can a specific spatial configuration afford one activity better than another? How do tools, props, scale, and access tend to shape activity, whether in meeting spaces, negotiation rooms, design studios, or public squares? Urban historians assume the existence of a city of collective memory, and thus the importance not only of its commemorative inscriptions but also of its casual conditions and their residues. Ethnographers seek evidence of unofficial "turfs" and tacit social geographies. Environmentalists ask which residues are considered pollution. Cultural critics look for connections between the urban pattern and the communication media politics behind it. Architects emphasize how the rise of infrastructures, such as electrification a century ago or smart green thermal systems now, creates opportunities and challenges for creative work. They ask how any of these serve collective memory, become a usable public infrastructure, or take on aspects of a commons.

The word *palimpsest*, a surface that shows traces of previous inscriptions, comes from the practice by medieval calligraphers of reusing precious sheepskin parchments. It was appropriated into urbanism by critics of clean-slate urban renewal. Better that the city bear traces of its past. Better that continual development occur bottom up, full of gaps and imperfections, than top down according to some totalizing quest for transparency.

For if all experience were as erasable as a screen, some longer views about place in the world might suffer, even vanish. Some authoritarian views could be imposed overnight. Individuals, interest groups, and institutions might lack the means to curate ever-changing, externally imposed phenomena. A new dark age could arise amid information superabundance. Instead of scarcity so extreme that you had to scrape used sheepskins to obtain any writing surfaces, under extreme superabundance just about everything could be an erasable writing surface. In both cases perhaps too few inscriptions would persist.

The perspective of architecture and urbanism may prove useful to work in cognitive science, interface arts, and the history of information. When several fields converge to gain a better grasp of situated attention, perhaps they will produce greater well-being. Perhaps they will know better whether persistent, inhabitable, inescapably shared information becomes a new kind of commons. Perhaps there is a topology to embodied information. Tagging a GPS coordinate with a label is not the same as covering a wall with an image or filling a space with a sound. Thus, for a start at the environmental history of information, a topology of points, areas, volumes, and networks forms the categories and sequence of the chapters to come (figure 5.5).

Topology	*Technology*	*New conditions*
Points	Tagging	Participation—new epigraphy
Areas	Display	Beyond cinematic perspective
Volumes	Buildings	Comfort is not uniformity
Networks	Urban computing	Infrastructures unbundled
Environments	Ubiquity	Information environmentalism

5.5 One topology of the ambient.

In architecture and urbanism, you can't just turn off the screen, or flip to another channel. You have to live in the artifacts. Despite how private and public players have shaped particular elements of built space, there is always some common circumstance to inhabitation, a circumstance that shapes action and perception. Now, as information technologies infuse these circumstances, how does that affect inhabitation? Is there anything about architecture's cognitive role that must be upheld or lost, lest it be covered in more ephemeral phenomena at lower resolution? Conversely, where mediation enlivens a space or makes an urban resource more usable, is there anything about its inhabitation that invites perception as a commons?

5.	FIXITY
Main idea:	Architecture's cognitive roles
Counterargument:	Increased mobility
Key terms:	Mise-en-scène, sensemaking
What has changed:	Pervasive information technology
Catalyst:	Out-of-context media
Related field:	Architecture
Open debate:	Value of persistence?

II Toward an Environmental History of Information

Tagging the Commons 6

What is an environmental history of information? The act of tagging the commons invites this question. Many kinds of markings reveal the city as a collective cultural work. They also trace the remarkable diversification of information technology. So in its explorations of the ambient, this inquiry now turns from the subjectivity of attention to the objectivity of information in the built environment, and, where possible, to its history.

This begins with the tag. There is no simpler piece of situated technology. Although overtly semantic itself, a tag quickly shifts attention to the intrinsic structure of whatever it labels. Tags are simpler and possibly more prevalent than screens, which are the assumed focus in today's economics of attention. Whereas the contents of a screen are disembodied and usually disengaged from context, a tag is almost always about something right here. More basically still, a tag is physically inscribed and not sent.

Once upon a time, the ancients stood stones in an enchanted landscape. Later on, emperors chiseled proclamations into marble facades. Opposing street gangs scrawled slogans on walls in the night. Shopkeepers washed down the sidewalks each morning, then put out menus of the day's wares. Barkers pushed broadsides, often literally in your face. Prefects and police posted notices and regulations. Commissions installed commemorative plaques. Soldiers and sailors proudly flew their colors. Portraits of Lenin and Stalin hung over apparatchiks' desks. Fresh batik draped over old statues of Ganesh. Crosses and crescents glinted in the sun atop spires. And more recently, bank logos glowed atop skyscrapers through the night. Advertising campaigns advanced onto all manner of surfaces that formerly lacked inscriptions. Teenagers spray painted tags around town, likewise building their brand.

Digital media now transform these tales. Tagging has turned technological, from huge programmable facades to wearable fashion accessories to tiny radio frequency identification (RFID) chips. A new kind of information commons, different from those in disembodied electronic cyberspace, may be taking form at street level.

Architects again admit of signage and ornament. Planners make policies about light pollution and mobile noise. Information designers build wayfinding systems for use on handheld gadgets. Interaction designers seek delight in embodied social navigation. These design and cultural opportunities may rival those of any past era of technological change, such as electrification a century ago. The cultural costs could well surpass those of other ages as well, even those of automobiles to cities.

Situated technologies may not dominate everyday conceptions of information technology beyond smartphones and screens. Surveillance usually draws more concern. Big data clouds now draw the most business hype. More basic concerns about digital divides in supposedly ubiquitous Internet access precede more abstract concerns about ambient information. In other words, the cultural shift from what was called "cyberspace" (a quaint word by now) to pervasive computing seems only partial. Obsession with mobile smartphones masks an equally important design challenge of embedded sensors, data, and processing. So let the simple tag stand for the fact that there are situated technologies, too.

Beginning from the simplicity of the tag, let this inquiry take a topological survey of the expanding contexts and formats of information that are so altering the nature of attention. Consider tags as points, glowing surfaces as areas, architectural space and atmospheres as volumes, and urban resources as networks. To explore how the ambient becomes a reality and its design and governance become cultural necessities, begin from a timeless, everyday, and utterly local act: writing on the walls.

Urban Markup Languages

The simple act of marking the city reveals larger cultural aspects of the commons. To explore such aspects, consider, for example, the simple sticker. More so than its meaner aerosol counterparts, adhesive art seeks some shared cultural ground.

Indeed, the adhesive tagger Shepard Fairey, creator of Obey Giant, of which some half million instances were once said to exist,[1] and of the memorable red-and-blue Barack Obama campaign poster, whose imitations and mockeries seemed to rival the Giant in their ubiquity, was for a time considered the most eminent visual artist in the United States. In an illustration of how motifs and tastemakers cross cultural borders, the Institute for Contemporary Art in Boston marked the occasion of the Obama inauguration with an exhibition on Fairey, and with a Giant on its facade (figure 6.1).

On the street, a tagger is someone who signs in aerosol. To tag is to spray paint your name. Most people consider aerosol signing to be antisocial, and many cities have made it a crime. Yet among insiders, aerosol signing provides what an information professional might call a "reputation system." Anyone who can sign all over town without getting caught in the act must be a "badass." At some animal level, in the realm of embodied cognition, tagging just marks territory, without civic aspiration. But then it becomes social—defiantly in its choices of site, competitively in which tags are respected and not soon written over by rivals, and culturally in how some signs become noticed by the general public and even appreciated by critics. Why else would so many art museum gift shops offer coffee-table catalogs of graffiti?

Graffiti.org, the most-linked compendium of curatorial-minded graffiti enthusiasts, called itself "Art Crimes." Washington's arbiters of art have said otherwise, however: "Large graffiti pieces are also on display in the 'hallowed halls' of the National Portrait Gallery."[2] Yet before curators official or unofficial step in, the corporate coolhunters have already been through. Tastemaking mines street culture first. It is difficult to make cultural generalizations about the street, where there are as many styles as there are taggers. For present purposes, it is fair to assume that conflation of art, crime, curation, hyperlinking, and online tagging of sampled street art only adds to how many ways, and to how many different sets of eyes, fresh markup stays hip. "I think we owe everything to the Internet," traffic sign modifier Dan Witz has observed.[3]

6.1 Obey Giant on the Institute of Contemporary Art (ICA) building, Boston, 2009. Photo: Joe C./www.random-pattern.com.

The aggregate of so many tags on the street is more problematic, however. Citywide, rampant graffiti indicate distress, and tend to invite other troubles. Anyone who remembers New York in the 1970s will know this to be true.[4] There the graffiti reached an unprecedented scale, most memorably covering almost all subway cars (figure 6.2). In response, subway rail yards got fenced in concertina wire, wide-nibbed markers were taken off the market, and penalties for aerosol signing began to escalate.[5]

As New York's recovery from its troubled conditions of the 1970s made clear, one of the most useful public policies is to fight any appearance of anomie. It is important to assert the existence of a commons. This position was made memorable by the theory of "broken windows," introduced in the early 1980s: "Social psychologists and police officers tend to agree that if a window in a building is broken and is left unrepaired, all the rest of the windows will soon be broken. . . . Vandalism can occur anywhere, once communal barriers—the sense of mutual regard and the obligations of civility—are lowered by actions that seem to signal that 'no one cares.' . . . Such an area is vulnerable to criminal invasion."[6]

Among the side effects of New York's cleanup, more interesting variants of tagging appeared. Adhesive art or "slap tagging," for instance, Shepard Fairey's medium, took off with "Hello My Name Is ________" stickers, normally used for casual business receptions. Today the "I Wish This Was ________" sticker uses the same visual format for suggesting street-level improvements.[7] Stickers had a past in posters, of course: long before electronic media, posting lithographic bills was the main form of advertising. Thus most cities have rules about flyposting. But compared to spraying, stickering makes it easier to hit more locations, and easier to stay out of jail.[8] Still less risky is "reverse graffiti," a new genre of erasure, which helps owners remove grime from their walls. Such noncriminality has helped move street art beyond teen angst and into the cultural mainstream, such as museums.[9] Stylistically, there is something in the air that favors urban markup.

The rise of electronic tagging may not hurt this. Many people and organizations of respectable means now have their

6.2 Graffiti on New York subway cars, 1973. Photo: Erik Calonious, EPA.

own use for the word. To tag is to insert keywords into content, for example, as when photo sharing in FlickR. Lately, the most fashionable format has been the hashtags of Twitter. More generally, tagging implies all manner of metadata: smaller identifiers for larger pieces of information. The openness is one reason why tagging has exploded in popularity. Anyone can make up a keyword, and there is no such thing as a wrong tag. And now with GPS location data, RFID identity tagging, and augmented city smartphone overlays like Layar catching on, a new middle ground of tagging has come into being.

The curatorial prospects seem huge, far beyond what museums have done with purely physical tagging. Bottom-up tagging online has great potential for emergent effects. As of 2012, the augmented reality business appeared ready to explode.[10] Whatever may happen, the complex patterns that arise from very large numbers of very simple elements may eventually stabilize into enduring relationships, and useful classification schemes.[11] That, in turn, provides an excuse for going out on the town.

Carved Inscriptions

Stickers don't last the way carved marble does. The arts of urban markup differ not only by intended effect, but also by duration. To "set it in stone" means to leave something for the ages. Stone still has clout, even in a media-saturated age. Of bygone eras when written communications were scarce, one can only imagine the power of a proclamation chiseled into a stele or portico.

The discipline of epigraphy studies the messages found in archaeological remains, usually in stone. This has long been important to classical studies. Universities had epigraphers before they had architects. Epigraphers must interpret findings against whatever else is known about a culture through other means in order to establish enough context to make sense of tags. Necessarily short tags "tend to omit pertinent information that is already known by the intended audience," observed epigrapher Bradley McClean,[12] who listed eighteen categories of stone markings commonly found in classical Mediterranean cultures (figure 6.3).

Architecture has often provided both physical and cultural context for making sense of inscriptions. Architectural historians assert how, long before print, the inscription and ornament of buildings provided an effective information medium. For example, the facade of a church could at once, through its iconography, educate the laity—perhaps also providing a pulpit from

1. Decrees, laws, treaties, official letters
2. Honorific decrees, proxeny decrees, and honorific inscriptions
3. Dedications and ex-votos
4. Prose and metrical funerary inscriptions
5. Manumission inscriptions
6. Other legal instruments of common law
7. Boundary stones
8. Milestones
9. Herms (pillars to Hermes)
10. Sacred laws
11. Other sacred inscriptions
12. Inscriptions on public and private works
13. Accounts and catalogs
14. Inscriptions on portable objects
15. Quarry and masons' marks
16. Inscriptions in metal
17. Graffiti
18. Artists' signatures

6.3 Categories in classical epigraphy (source: Bradley McLean.) Background: stele, with decree of phoros (tribute) to Delian League (447 BCE). Photo: Marie-Lan Nguyen/Wikimedia Commons.

which the clergy might speak—and, through its magnificence, stand as a visual sign of the aspirations of the society that built it. Of course, cultures differed in how their sacred structures performed this didactic role, whether through tags and texts, images and imposing facades, or meaningful sequences of spaces; some structures were fairly covered in script. It is an oft-cited story in the history of architecture how print usurped this role. "Architecture is the great book of humanity," Victor Hugo famously lamented on the rise of print, "Gutenberg's letters of lead are about to supersede Orpheus's letters of stone."[13]

Although the Victorian city was full of everyday print uses like none before it,[14] the arts of inscribing stone continued, and even increased for a time, especially with respect to architectural ornament. Steam-powered industrial machines made the rough work much easier, which gave skilled stone carvers time for more of the projects that slowly changing cultural tastes still demanded. Medallions, swags, and sgraffites ornamented holy and unholy edifices alike (figure 6.4). This produced an overload in its own right: fine ornamental motifs were executed more often (now outside of traditional cultures) just because it had become technically feasible. As crude new technologies served ornate tastes still shaped by previous handicraft sensibilities, the results were especially ponderous. Not only in buildings, but also in clothing and the decorative arts, overdone ornament became the signature of the Victorian age, at least as seen in retrospect from the twentieth century, after skills and tastes had changed.

Today, a single gravestone, one of the few remaining instances of tagging for the ages, can cost thousands. To fund the carving of a simple dedication in the cornice of a building takes the resources of a state or, at least, a grand public institution. The artisans who carved stone for the Beaux-Arts libraries, concert halls, and museums of the industrial city aren't so numerous now. In the recent past, you could find a few at work on the Cathedral of Saint John the Divine, in New York, where they were supported by diamond-edge digital gang saws that did

6.4 Trade inscription for an architect's office, Paris.

all but the finish work; more recent advances in robotic water jet cutting are renewing the art of stone cutting, but only in a handful of studio cultures.[15]

Official, exclusive, and enduring, stone is graffiti's pure opposite. Except the two do share one trait: relationship to a commons.

Rampant Signage

Most official inscription today supports what matters most: moving about safely. To alter traffic and safety instructions on your own could be just as much of a crime as aerosol signing.

Standardized instructional signage didn't always exist. Consider its rise in Paris, for example. In her influential history of the city as text, Priscilla Ferguson explained the early nineteenth-century program of naming and charting the whole of Paris: "That streets should have names is not self-evident. For centuries, most villages and towns felt no need to name their streets, and even today a major urban center like Tokyo manages to do without them."[16] This naming program was a major reconceptualization. "Street names and other symbols, [Abbé] Gregoire reminded the Assemblé Nationale, provided the Revolution with means to do what no regime had ever done—institute reason and popular sovereignty."[17] Contrary to popular misconceptions about their neutrality, tagging systems exist with a purpose, by and for particular constituencies. Not everything gets tagged, and systems and selections of tags define and shape the groups they serve. Writing about the rise of guidebooks, but with implications for other uses as well, Ferguson observed how publishers felt that "unmediated contact with the city is inadequate at best, and probably dangerous as well."[18]

Today a small industry in environmental graphics serves these needs. In America, the Society of Environmental Graphic Design (SEGD) claims 1,600 members, who work on "wayfinding systems, architectural graphics, signage, exhibit design, identity graphics, dynamic environments, civic design, pictogram design, retail and store design, mapping, and themed environments."[19] Yet members of many other disciplines have also engaged in these pursuits. Environmental graphics was a growth industry long before digital location tagging became widespread. Just look at all the signs going up lately to proclaim what once was obvious to all. "If the door is closed, do not enter," one sign parody reminded, at the 2009 Edinburgh Festival Fringe.[20] Operational and safety instructions nearly rival advertisements in their ubiquity.

Instruction occasionally operates with intrinsic structure instead of placard signage. For example, California color codes its curbs in paint to indicate parking restrictions. In Germany, differences in surface texture remind bicyclists and pedestrians which band of the sidewalk to use.

Officialdom sometimes even operates in aerosol. Thus construction crews spray color-coded arrows on the pavement to indicate utility lines below. When the phrase "urban markup" was catching on among digerati, who would (unofficially) "war-chalk" free Wi-Fi hotspots for one another's benefit, *Wired* magazine ran a piece on pavement spray tags:

> If you know the lingo, you can visualize the dense architecture that sprawls beneath our streets. The paint colors are fairly standard: Red denotes power lines; yellow flags oil and gas; blue is for fresh water; green indicates sewage; purple highlights reclaimed H_2O; and orange tags communications or cable TV lines. Some acronyms, like MCI or SBC, are obvious; others aren't. IP, for example, means "iron pipe," and U.S.A. stands for "underground service alert" (the aforementioned area slated for excavation). As Mike Hart, a plumber for the San Francisco water department puts it, "I tell my kids that I'm a graffiti professional."[21]

Meanwhile, whether with logo or image, advertising tags most widely. This too has a history. In the tradition of architecture as display medium, especially before the rise of broadcast media, businesses painted advertisements on the sides of their

6.5 Hand-painted signage, Beaufort, South Carolina. Photo: Walker Evans, 1936 (Minneapolis Institute of Arts).

buildings (figure 6.5). In the nineteenth century, color lithography pushed postering into widespread public view, which led to restrictions on flyposting resembling those on graffiti. Then, in the twentieth century, electrification blurred distinctions between broadcast and architecture in new, and often inescapable ways. Today, electronic overlays accelerate and transform that process. Advertising tags locations, consumers, and transactions alike, and then aggregates and mines them. Advertising makes brands into places and places into brands. Even more so than with image, advertising operates in trademarks, logos, and other point-specific, one-dimensional signifiers of brand. Moreover, it does so transgressively, like graffiti, ever advancing into places formerly free of its tags.

In his huge 2003 volume of street-level photography, Tokyo-phile editor Eric Sadin called it "Times of the Signs":

> Giant screens. . . Printed matter. . . Information facades. . . Interactive terminals. . . Screen arrangements. . . Flags . . . Billboards. . . T-shirts. . . On the tips of our toes (walking on characters). . . Global brands. . . URLs in the city. . . Mobile phones. . . Luxury brands and architecture. . . Video games. . . Signage. . . Neon signs. . . Pachinko. . . Print club. . . Karaoke. . . Sex stickers. . . Information [sandwich] men. . . Public phones . . . Media buildings. . . E-learning. . . Electronic billboards. . . Sound information. . . Surveillance. . . GPS. . . Internet cafés. . . [22]

Cases in Adhesive Electronics

As a way to rethink handheld urban computing, location-based media, and their interplay with electrified architecture, you might start with simple pixel liberation. Not every square of light needs a frame. Anything that lights up and can be attached to some other surface potentially becomes a tagging system. If it also communicates, it can become part of a larger image. One serendipitous such device, which caught the imagination of artists and designers online, was the "LED throwie," developed by Graffiti Research Lab in 2006. Each consisting of one or more LEDs (light-emitting diodes), a coin battery, and a rare-earth magnet, LED throwies, as their creators winked, were "an inexpensive way to add color to any ferromagnetic surface in your neighborhood."[23]

Adhesive art shows how location-based media don't always need GPS or Bluetooth to be known. It raises an important question about environmental awareness and the ambient. Can the purpose of handheld electronic media move beyond communicating for the sake of communicating, beyond tuning out so much of the world through personalizing everything, to helping someone be here now, in the sense of knowing an urban commons?

An especially well known project for such purposes was Yellow Arrow (2003–07; figure 6.6), recognized by the *New York Times* as one of the earliest instances of "the Internet overlaid" on the physical world "to make the city more browsable,"[24] and exhibited by New York's Museum of Modern Art in 2008. The arrow itself took the form of a palm-size sticker. Each sticker had a unique alphanumeric code to use in text messaging to and from Yellow Arrow's servers; you could buy one for 50 cents, stick it anywhere you dared (or had permission to, as the organization advised, to keep it legal), and upload a short text comment about that place. Passers-by who came across the sticker could then text the indicated code and read your comment or upload another of their own. Over the four-year run of the project, several thousand stickers were applied, mostly in a few pilot cities. Even though most arrows were one-offs (as is natural in such a bottom-up authorship format), some civic themes did emerge. In Copenhagen, the arrows were often used for political debate, for example about infrastructure issues, and in Boston, mainly to serve the cause of bicyclists' rights.

In hindsight (Yellow Arrow closed for reasons of scale and funding), much of the project's appeal came from the interaction aesthetic of physical tagging. Whereas the social networking

6.6 Yellow Arrow (2003–08), some instances in Copenhagen, tagging transit data. Photos: courtesy of Jesse Shapin.

aspect may have prevailed at the time because Yellow Arrow was seen as a community by some participants and as a way of life by a crazy few, more recent technologies and especially Twitter have taken that experience to new and different levels. Instead, the physical placement aspects of the project stand out. Here was social networking that was not ubiquitous, that involved the delight of discovery, that tested the cultural and material constraints of the city, and that tapped into the unofficial, uncriminal coolness of stickering. "It's been called a game, a form of graffiti, and the largest performance art piece ever attempted," wrote the *Harvard Crimson*.[25] Yellow Arrow was something you could do on a dare.

There have been even simpler early web-served tagging systems. Grafedia first took open-keyword media sharing to the walls. Its assumed medium was handheld picture messaging, although it also allowed small texts and sounds. Having invented a keyword and posted some annotations with it, a user would tag in blue chalk to resemble hyperlinked words on the web, and so engage literally and physically in hypertext markup language. This has not become widespread.

Far more frequent is tagging a URL (Universal Resource Locator, or web address) with a QRC (Quick Response Code) sticker, which is, in effect, the two dimensional barcode. Indeed, in Japan, where it shows up everywhere from food packages to gravestones, the QRC format is well on its way to supplanting the UPC one. Worldwide, the best-known early instance of QRC tagging in the geospatial web was Semapedia, which linked to Wikipedia entries. Semapedia generated a smartcode file for any Wikipedia article, and sent that file for printout, for use in tagging. Launched in 2005, Semapedia generated over 50,000 tags. This is an obvious early instance in curating unofficial markup. Encyclopedias lack street appeal, however, and one like Semapedia could be appropriated by anyone generating a smartcode for some other site, although a Google search for "Semapedia abuse" still yielded a null result in 2012.

Instead, recent competition in applied electronic tagging tends to focus on RFID (radio frequency identification), with its lucrative advantages in supply chain management, especially amid the huge growth of merchandise logistics in China.[26] Thus, in the spring of 2008, Walmart began charging its suppliers a small fine for each palette of goods not tagged with RFID. On a smaller scale, Avery Dennison, the maker of the now historically significant "Hello, My Name Is________" sticker, now also operates in "RFID Labeling Solutions."[27]

Social thinkers rightly shun RFID for how it reifies everyday data surveillance. With RFID on your passport or transit pass, no longer do officials have to ask you for some ID; they can just scan for it. When coupled with data mining techniques, data surveillance leads to socially corrosive behavioral tactics, whether on the part of marketers, police, or interpersonal networks. For many years, esteemed futurist Bruce Sterling kept an "arphid watch," which cataloged the scary realities of tracking living things, possibly even errant humans, in some cases, by means of implanted Verichips.[28]

For a taste of social fiction, in which RFID plays a more innocent role at tagging, it might help to take a moment to question that most widely assumed social act: web posting. Instead, imagine if there were nothing server-side. For instance, a system of RFID tags, local flash memory, and off-the-shelf reader bracelets could house a threaded urban narrative on a specific theme, say ethnic history, urban botany, or architectural preservation, and, with nothing posted centrally, data surveillance would be less likely, paranoia might abate, and participation could become more candid. To read a different topic, you would buy a different reader off the rack at a local store. That does seem difficult to imagine. Naturally, the temptation would exist to gather all this lore, online in a box on a server farm outside town. Where does that temptation come from? Would there ever be counterexamples? Under what circumstances of noncriminal, nonofficial tagging would server-side posting become taboo?

Many technologists are now working on a more viable development: "augmented reality" (AR). This expression has been around for at least twenty years, commercially viable products began to appear in the last five, and the prospects for explosive growth seem quite serious. The basic idea is certainly one of tagging. Different technologies determine how those tags overlay onto the visual field. The simplest way combines tagging with the use of orientation chips (GPS, compass, plus accelerometer) now included in smartphones. When introduced on the first Android devices in 2008, this combination led to a first wave of augmented reality platforms, such as Layar and Wikitude, which popularized the expression "reality browser," and which began what soon became a flood of local apps. Unlike the use-anywhere apps that have proliferated over smartphone networks, these local apps tend to be use-someplace ones, and can therefore be even more numerous. Right now, this new medium is changing too quickly for a print publication to follow (just search "augmented reality apps" next time you need a guide in an unfamiliar town).

To achieve the spirit of frameless displays and palpable augmentation takes much better tracking and image registration that holding your smartphone up to a city scene can provide. The technology has been around in military and emergency response operations in the form of heads-up displays on windshields or eyeglasses. As of 2011, Google announced plans to introduce a fashionable consumer device in the form of wraparound eyeshades. (Many a technofuturist believes the handheld smartphone is just a clumsy stage on the way to better wearables.) With respect to visual attention, heads-up and registration both mean not having to look away from the scene for the annotation—no looking a few degrees downward at your smartphone, for example. Tracking means that the alignment follows your gaze quickly enough. This tends to reinforce a bias toward attention as selective visual gaze. If, on the other hand, the overlay becomes too immersive, as happened in many head-mounted goggles of the past, the disconnect between visual saturation and embodied systems of haptic orientation will quickly produce simulation sickness. This is not so simple as walking around with an iPod.

Local content for augmented reality overlays quickly followed themes from more conventional guidebook technologies: local restaurants, histories of local landmarks, and thematic walks to take. Architects took note of the prospect to overlay photographs of past and future conditions onto current views of the city. In 2011, MetaIO, a pioneering AR company, introduced a three-dimensional rendering technology for building urban augmented reality applications, such as for showing past or future buildings in context. Elsewhere, most of the early work has used FlickR images. Compared to other things to do with a smartphone, these seemed like ways to tune in rather than out.

With such prospects in mind, this inquiry next moves from positional tags to image overlays. To cover the world in electronic images may be neither possible nor desirable. To fill heads-up fashion eyewear with such images (at least where able to do so without causing sensory-motor disconnect) has the advantage that fewer messages might crowd physical space to be seen by everyone.

Location-based media in varying degrees of augmenting reality have grown from a curiosity to a big business. Geodata support enterprises in environment, infrastructure, logistics, social services, security, and more. The geospatial web increasingly collects and delivers these managed data on demand, often to the very places they document. Embodied computation gathers and distributes these feeds in uncanny ways. Design conferences such as Lift and Where 2.0 feature the latest blogjects, mashups, and distributed narrative installations. Research societies focus on specific technologies such as positioning, sensing, embedding, displaying, and ad hoc networking, among many others; some even defend the electronic commons.

Many of these rapidly sharing domains share assumptions about ubiquity and mobility that raise deeper cultural concerns, chiefly about privacy and surveillance, but about many other things too. For instance, how does so much personalization recast citizenship or civility? What kinds of information best belong in one place and not everywhere? How might augmented reality media help document and conserve material and energy flows? Does ambient connectivity enhance or distract from environmental awareness and participation? The field of urban computing has emerged to explore these concerns.

Because the adoption rate of mobile handheld communication surpasses that of cars or television, it is rightly considered the most transformative urban infrastructure of the day. Mobile applications of the geospatial web—not always for positional wayfinding as geodata industries too quickly assume—thus tend to dominate artistic and academic investigations. Agendas in social navigation and environmental management also emerge. A bottom-up surveillance (sousveillance) campaign may flag locations where surveillance cameras are active. More ambitious sousveillance may try to reverse corporate greenwashing by exposing cultural offenses.

Tagging thus crosses from personal territorial marking into electronic art and technological design. When a large new field grows up around a given focus, some of its smaller, less dominant aspects may help shed light on its new outlooks. For example, to question ubiquity, it helps to study the situated.[29] To question mobility, it helps to study fixity, to look at, say, the expressive urban material constraints that rude young aerosol tagging ignores. And thus also, to question the spellbinding high-tech complexity of handheld social networking, it helps to consider low-tech and sometimes antisocial tagging.

Awareness of a Commons

In the literature of the commons, there exists a well-known trope on tagging, "the 'I' and the 'it,'" first advanced by urban sociologist Richard Sennett in his 1990 book, *The Conscience of the Eye*. Does a youngster spray painting a subway car in the Bronx see that subway car? Sennett recalled New York in the 1970s: "The scale of this graffiti was what made the first impression: there was so much of it. . . . The kids were indifferent, however, to the general public, playing to themselves, ignoring the presence of other people using or enclosed in their space. . . . Transgression and indifference to others appeared joined in these simple smears of self, and with a simple result. The graffiti were treated from the first as a crime."[30] Whereas a tagger just shouts "I," the "it" expresses the presence of others, past, present, and future through the material forms and constraints of the city. There are limits to personalization. Working with external circumstances that result from the presence of others pushes you to a higher level, and that is an important aspect of urbanism. The aggregate of these material expressions constrains

6.7 Public service: line chalking as tagging. Eve Mosher, HighWaterLine, 2007. Photo: Hose Cedeno.

each individual contribution. To Sennett, whose more recent work on craft affirms this outlook, an artist working in a civic capacity does indeed see this material commons and lets it shape his or her intentions and expressions, through which the city becomes a medium.

To sense a cultural accumulation can be the first step toward recognizing a commons. The mature tagger can see the city as the cumulative state of many people presenting themselves to one another. Acts of tagging can add to the understanding of the city as commons, rather than detract from it. For example, in an early instance of public awareness graffiti and ambient ecofeedback, Eve Mosher's HighWaterLine (2007; figure 6.7) chalked

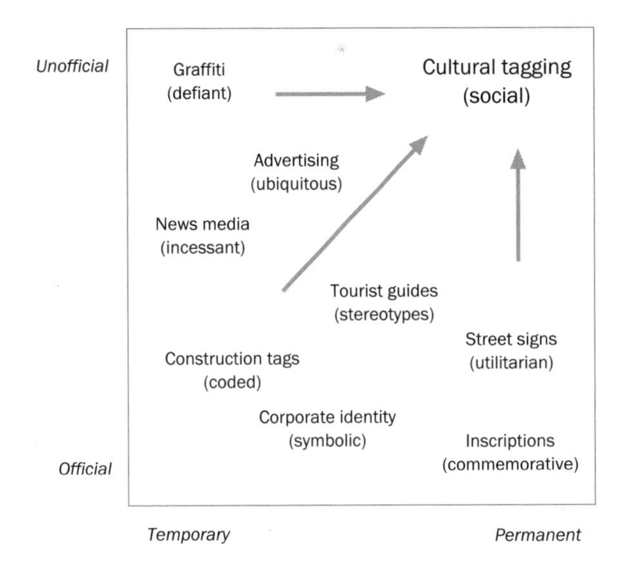

6.8 Toward a new epigraphy.

the 10-feet-above-sea-level contour through the most densely built parts of New York. No degree of augmented reality tagging has quite the impact of old-fashioned chalking.

Ideas of the commons often resurface in this inquiry. We must return to them with respect to networked urban resources, with respect to environmental history of information, and with respect to attention itself. They raise fundamental questions about civility, the distracted urban citizen, and the public good.

Between the age-old extremes of transgressive graffiti and official proclamations, urban markup forms a new middle ground (figure 6.8). More persistent, interlinked, and filterable than graffiti, but more improvisatory, narrowly themed, and socially engaging than official signage, the tags now appearing in location-based media invite consideration as a genre in themselves. Do they invite a new epigraphy? What does electronic tagging do for the upkeep and valuation of physical commons?

6.	TAGGING THE COMMONS
Main idea:	New cultural middle ground, between official inscriptions and transgressive graffiti
Counterargument:	Semiotics has been there, done that
Key terms:	Tagging, epigraphy
What has changed:	Tags can be filtered and linked
Catalyst:	Rise of location-based media
Related field:	Environmental graphics/wayshowing
Open debate:	Will anyone curate urban tags?

Frames and Facades 7

Everyone knows where attention goes: to all those glowing rectangles. So where a general history of information might explore how digital displays have superseded print, an environmental history of information explores how they have proliferated in physical space, as when screens cover whole building facades. Digital displays now number in the billions. According to Displaybank, already nearly half a billion large-area TFT LCD (thin-film transistor liquid crystal display) panels ship from east Asia in a single year, as do more than a billion small-format LED panels for cameras and smartphones.[1] Then, besides the billion or more smartphones being carried around, billions of other displays are being built into everyday life: at points of sale, outside meeting rooms, on parking meters; as electronic paper, data murals, electronic billboards, huge media facades. At a baseball park, the crowd watches the larger-than-life close-ups of itself on the JumboTron, more than it watches the game. At street level, a hyperlocal events site such as EveryBlock takes

7.1 Bulletin boards at street level: *Boston Globe* war map, 1944 (Getty Images).

photo sharing from the web to the sidewalk, supplanting the now obsolete newspaper and recalling the town bulletin boards of former times, which existed well into the twentieth century (figure 7.1).

Picture a car pulling in for some gas. Both the car and the gas pump have a video screen, where ten years ago they did not. The kids in the backseat have screens of their own, as if the world going by is no longer worth watching. A kiosk inside the express mart has several touch screens for the sole purpose of selling lottery tickets. This frees the attendant to watch a bank of security monitors instead. A television, placed up high atop the coolers next to the Buddha figurine (as if both sentimental totems of past cultural tradition), is always on though seldom watched.[2] The widescreen display above the snack rack gets more attention: larger, more visible from the entrance, and programmed with much higher production values, it presents a constant stream of mostly close-up, slow zoom-in shots of food.

Today fewer screens require you to sit down or to fix your gaze for more than a moment; fewer fill your field of view. Multiple screens may compete for your attention, or they may recede into a background of possibilities for shifting attention. The contents of their fragmented displays change across a greater range of time periods and in response to a greater range of circumstances than was possible before. Not all displays describe someplace else; sensing, networking, and embedded computing increase the capacity for displayed images to be about current conditions in their immediate surroundings.

The ambient emerges from this visual abundance. In this it has reached a new stage. Vision famously fragmented in the twentieth century, from cubist painting at its start to clickable windows by its end, yet it mostly kept its frame. Indeed, you could identify that century as the one where people sat down passively in front of framed, flickering screens. Now, as display technology diversifies in size, role, and use, visual culture is accelerating and transforming once again. The more that images diversify, proliferate, and compete, the less any one of them may succeed at capturing your attention. Instead, they all fuse into a landscape, in which the perspective furnished by any one frame yields to a new kind of perspective on a world full of them. These new display practices erode assumptions about the cinematic nature of the frame, and instead belong to the world of architecture.

The Facade Communicates

This inquiry into the history of information began from ancient inscriptions, many of those on buildings. In a way, buildings were the first mass communication medium. Icons in architecture served to teach an illiterate laity (figure 7.2). In much the same way, textual inscriptions in architecture served to instruct a literate public. These weren't just tags: although many acted as titles, narratives, schedules, rolls, and proclamations, they also had intrinsic spatial attributes in the design, layout, and placement of their characters. This was especially so in cultures whose prohibitions on figurative art led to expressive outlets in calligraphy, which itself had spatial attributes. Many of these communicative elements were intrinsic to facades. Picture a ruler speaking to the populace from the portico of a palace. In Istanbul, the High Gate of the Topkapi Palace, from which the Ottoman sultans and their delegates spoke, became synonymous with their imperial power; both became simply known as the *Porte*.

The words *edify* and *edifice* relate. Murals, illustrative ornament, and form itself all made architecture communicate. Between the persistence of its artifacts and the relative lack of other media, architecture imposed ideas with a power that would now be difficult to imagine. Too few histories of information acknowledge this architectural power; too few histories of information are environmental.

Even a history of writing can become a history of certain spatial dispositions.[3] Buildings instructed not only directly with text and images, but also indirectly through the devotion implied in their workmanship and their permanent embodiment of messages. The extensive embellishments of a French medieval church would have meant less if fabricated with machines rather

7.2 Facade that instructs: portal to medieval Church of Saint Trophime, Arles (twelfth century). Photo: Steffen Heilfort/Wikimedia Commons.

than by hand. Thus, in many stages of a building's history, distinctions between text, images, and ornament can seem to all but disappear.

How buildings have been transformed but never made obsolete by successions of modern media, whether print, broadcast, or digital, remains a perennial theme in architectural history and theory. Now, as displays ranging from tiny touch screens to huge media facades provoke new kinds of controversy, many more disciplines beyond architecture have come to appreciate those histories.

Architectural form can, of course, communicate without any inscription. Thus a tripartite arch can stand for a particular military victory. Thus, too, the massive, imposing facade of a building suggests permanence, and the spire of a church, ascendance. The great stupa of Borobudur, ornamented with some 1,400 carved stone relief panels of scenes from the Sudhana, symbolizes the journey within, to be acted out by ascending its spiral pathway. Through form and not just annotation, architecture represents organizations to their constituencies. In the process, it also tends to represent the cultural circumstances from which it arises.

Furthermore as postmodern critics took such delight in reciting, formal signifiers have developed a visual code of their own, sometimes independently of their referents. The meaning of a sign comes as much from the circumstances of its creation or its reading as from what it refers to. In the free play of signifiers that results, the forest of signs becomes a delight in itself.

No wonder most buildings simply serve as carriers of overt, literal signage. This was true even before modern billboard technology and electric lighting. The brick side walls of many nineteenth-century buildings were covered with hand-painted signs. Although you can see relatively little evidence of signage in surviving portrayals of urban scenes before modern, literate times, such as Italian Renaissance paintings, other emblems of trades, perhaps considered inappropriate to include in precious paintings, appear in the engravings of William Hogarth and other artists of the eighteenth century. A major obstacle to compiling a complete environmental history of information is how little of everyday streetscapes survives either in images or in print.

Glowing Forms

One chapter in a longer such history could dwell on the introduction and impact of electricity. With electrification, walls were not only written on, but lit up as well. Fire, the medium of ancient signal technologies, has long been used to embellish or illuminate cities, at least in part. But the light it cast didn't become integral to the urban scene until the advent of its much safer, brighter, more evenly distributed successor, incandescent lighting.

Famously, at Chicago's World Columbian Exposition of 1893, when direct memories of the Great Chicago Fire of 1871 were still quite fresh, the nighttime illuminations drew the largest crowds. Not only was lighting no longer coupled with the threat of conflagration, but it was now also fast, relatively cool, and unprecedentedly controllable.[4]

In the Jazz Age heyday that followed the spread of electrification, urban nighttime imagery became a distinguishing cultural phenomenon. Many American cities touted their own versions of New York's luminous Great White Way. At street level, the signature medium became neon. Amid the financial speculative boom and bust of the New York 1920s, one of the more easily traced historical threads is the rise and fall of the Claude Neon Company, whose stock took a path that today would look familiar to investors in many a 1990s dotcom. The crackle and glow of the neon tubes were attractors in themselves. The medium's lurid quality fit with the anomie of the age. As a harbinger of the ambient, it communicated without need for message. "What, in the end, makes advertising so superior to criticism?" Walter Benjamin famously quipped, "Not what the moving red neon sign says, but the fiery pool reflecting it in the asphalt."[5]

Architectural landmarks, which had always anchored mental images of the city, now appeared in what was literally a new light. Lithographic prints that were posted on so many walls could now be highlighted for emphasis and for visibility from greater distances. Thus the billboard industry took new form. New kinds of signs and facades integrated light in ways not possible with flames. Buildings became beacons (figure 7.3), and skylines became more sharply defined by night than by day. Some of these beacons displayed weather data, like those of Pittsburgh's Gulf Tower, for example: red if warming, blue if cooling, steady if clear, blinking if snow or rain.[6]

Later, backlit corporate logos assumed such prominence that nighttime skylines came to resemble groves of sign pylons. In some commercial landscapes, each of the seemingly pixilated features was itself a branded backlit sign. In the 1990s, star science fiction novelist Neal Stephenson named this phenomenon "loglo." "The loglo . . . is a body of electric light made of innumerable cells. . . . Despite their efforts to stand out, they are all smeared together, especially at one hundred and twenty kilometers per hour."[7]

Glow stands out best against a large dark landscape. Starting in 1936, Las Vegas pumped cheap hydroelectric power from Hoover Dam into a new kind of illumination spectacle amid the nighttime darkness of the Mojave Desert, a spectacle that eventually became America's largest tourist destination. The 1995 opening of the Fremont Street Experience pioneered the use of lights as pixels in large outdoor display surfaces, in this case, over the downtown neon belt already long known as "glitter gulch." Alas, this didn't endure long enough to earn historic preservation status. In 2004, a 12.5-megapixel LED array replaced the original incandescent grid.[8]

Today, many cities have designated signage districts, where advertisers are free to turn up the volume.[9] Especially in the boom of urbanizing China, the greater the range of things emitting light in your field of view, the more uncanny the visual effect. Amid the anomie of a Hong Kong frontage road, electronic billboards take root. For a period in 2009, a four-story full-motion image of talking lips played in one such frame.

Not only in signage districts but also through everyday design neglect in office towers and parking lots, so much light gets spilled that many cities now recognize and regulate light pollution. Some neighborhoods even restrict the use of the lowly boxed backlit sign.[10] This is a topic worth taking up later, as an instance of governing the ambient. But, for now, consider one extreme genre of digital display: the electronic billboard.

Little seems ambient about a fifty-foot LED image. Few electronic billboards go up for civic reasons. Few invite the kinds of imperceptibly slow patterning that could make so prominent a surface pleasant to live with. Instead, today's electronic billboards rival texting while driving as exemplars of dangerous distraction. Studies indicate sharply greater safety risks when a driver's gaze goes off the road for more than two seconds. As a matter of policy, many cities have banned full-motion billboards. Many have compromised by allowing electronic billboards that only flip fixed images. Thus California requires four seconds between frames, and many states require eight. Yet each such change creates a visual event that can be seen from a distance of ten miles or more.[11]

São Paulo, a city seldom cited for its visual harmony, banned billboards outright in 2007, at least in particular zones that became the direct opposite of signage districts. Between 2007

7.3 Building as beacon: Metropolitan Life tower, New York City, 1905. Photo: Jessie Tarbox Beals (Schlesinger Library, Harvard).

7.4 Noteworthy recent large-scale media facade: King's Road Tower, Jeddah, Saudi Arabia (CITILED). Photo: courtesy of Alberto Ramon, CITILED.

and 2009, nearly 20,000 billboards were taken down. The global advertising industry responded with alarm at this threat. The ban, though under reconsideration, was still holding as of 2012.

Los Angeles, the capital of cinema, surely loves a lit screen. Although famous for hilltop views of its grid of glowing avenues receding to the horizon, the city emits far fewer lumens per capita than Las Vegas and has, as yet, no signage districts on a par with Tokyo's. Nevertheless, having served as inspiration and site for that infamously prescient work of glowing dystopia, *Blade Runner* (1982), Los Angeles is now a poster child for billboard regulation. As thousands of billboards convert from printed vinyl to LED arrays, neighborhood associations and city council members have taken exception. In 2009, the city called a moratorium on new billboards. KCET, a local viewer-supported television station ran an exposé, "Billboard Confidential," that identified Los Angeles as the "illegal billboard capital of the world," and that explained the process by which advertisers silenced politicians by providing copious free advertising.[12] By the time slowly enforced regulations required the owners to remove a billboard, they had restarted the process in another part of town.

Media Facades

What happens when huge electronic displays become a persistent part of physical architecture? Jakarta has claimed what as of 2009 was the tallest façade to be programmably illuminated, for the Grand Indonesia Tower at fifty-seven stories; as well as the longest (as of 2012) LED screen yet realized, in the Taman Anggrek mall at over eleven hundred feet.[13] As of 2012, the tallest realized full LED display in the Middle East was on the King's

Road Tower in Jeddah, Saudi Arabia, at twenty-six stories high (figure 7.4), and imaginatively programmed with a variety of artistic, civic and commercial images.[14] From projects such as these, it seems clear that media facades, the big attention seekers of the electronic world, belong to an architecture for the age of YouTube. Instead of designing for still photographs in glossy magazines, the owners and architects of the twenty-first-century design for one-minute video clips going viral on the Internet.

Media facades became especially noteworthy around the time of the 2008 Beijing Olympics, whose features included the glowing Voronoi tessellation of the natatorium, and the LED proof of concept GreenPix Zero Energy Wall (figure 7.5a). LEDs integrate well with tiled skins such as the recently completed Iluma building in Singapore (figure 7.5b), which combines display with a breathable mesh of polycarbonate polygons, overlaid outside a conventional structural facade. This escapes the rectangular frame. The media facade artists of realities:united explained how they emphasize that any particular building is "not a monitor" (not that one anyway).[15]

Media facade technologies have certainly been advancing, most notably in transparency, which besides letting the inhabitants see out, also allows the form of the facade and its screens to remain visible when the display is turned off (or it eventually wears out). In the years just before 2010, the first widely used technology for transparent media facades was MediaMesh, which originated in Germany.[16] Integrating rows of LEDs into latticed metal tubes mounted like a sunshade over the structure of a facade (figure 7.5c), MediaMesh is a screen in the older architectural sense of that word, and not the prevailing cinematic or digital norm. Taking the form of a lattice rather than a panel,

7.5 Media façade surfaces.

(a) GreenPix Zero Energy Wall, Beijing, Simone Giostra & Partners, 2008. Photo: courtesy of Simone Giostra.

(b) Crystal Mesh façade, Iluma center, Singapore, realities:united, 2009. Photo: courtesy of Jan Edler.

(c) MediaMesh technology, up close. Photo: courtesy of AG4.

it combines material and computational elements in a refreshingly simple way. In perhaps its most widely known early application, in Milan, for the 2007 renovations of the Arengario Museum on the Piazza del Duomo, MediaMesh encased the scaffolding that regularly surrounds European cultural monuments during maintenance (figure 7.6), displaying supersized images of supermodels from the world of fashion advertising. The Milan installation thus drew a clear contrast between architectural embellishment and framed display.

Scale in space affects how an image reads: it reads differently on a huge scale, a personal scale, or a tiny, postage-stamp scale. An image that is larger than your body, and perhaps too large to see whole, reads differently than one you can hold in your hand. Huge scale makes close-up images uncanny: the smaller the object and closer the shot, the stranger the effect when displayed very large. The effect draws focus onto the object instead of its scene, and further reduces your awareness of the frame. Thus, instead of suspension of disbelief, which sitting before a framed perspective always involves, the effect is "out of scene," or disorienting.

Scale in time also affects how an image reads. A transitory image reads differently than an enduring one. An image that changes continuously with a natural cycle, like the shadows crossing a facade over the course of an afternoon, reads differently than one that imposes its own orchestrations of time and transitional effects. The ambient display, no longer constrained by the short time frames of broadcast, need not jump around like television programming, and could, for instance, operate on a weeklong, barely perceptible visual cycle. Building facades have the opportunity to do things over days, weeks, and years that

7.6 An early MediaMesh installation, Milan, 2007 (AG4 + GKD). Photo: courtesy of GKD.

YouTube clips, television ads, and drive-by electronic billboards cannot.

So despite the huge range of controversy surrounding media facades, the main argument seems simple enough. Media facades should behave differently than cinematic frames, and over longer time cycles than most video productions. They may sometimes work best without pictorial images. Because almost any new technology is used to do the same old things at first, the owners of media facades mostly treat them like billboards, museum banners, or video screens, in other words as frames for other genres of visual communication, and (so far) too seldom as genuine extensions of architecture.

Window, Screen, Frame, Facade

An inquiry into attention and an environmental history of information share a common interest in apertures. A frame sets its contents apart from their surroundings, to be viewed through an opening in some different way, often in some different perspective, as if through a lens; from one space to another, as if through a window; and perhaps with some filtration of light, as if through a screen. Consider these openings as attention devices in architecture.

In architecture, a simple, well-placed opening frames a view, regulates the flow of sound, light, and air, and dramatizes passage from one space to another. As Le Corbusier famously pointed out, the window is one of the best themes by which to approach the history of architecture.[17] The most compelling windows connect contrasting worlds: an office and garden, an operations booth and factory floor, a prison inmate and visiting spouse, a warm bedroom and a winter storm. Thus an English window seat provides a way to be both indoors and outdoors in rare fair weather in a climate that discourages the building of patios and loggias. Where the spaces on two sides of an opening differ in appearance, conduct, privilege, ownership, or atmosphere, the physical portal remains the most vivid framed threshold of all. By contrast, there is little so regrettable about modern building as the loss of windows that open.

The screen means something different, and often more active, than the frame. A screen (often in the form of shutters or blinds) can be quietly gratifying to reconfigure on demand, and across daily and seasonal cycles. Classical Chinese architecture made remarkably effective use of open lattice screens, for example. Today a smart green building skin adaptively responds to changes in the weather. Of course, a facade might also serve as a video display device, a screen in the more usual sense of the word. Imagine the word *screen* as a verb: in more general use, to screen can be to showcase, to shield, to filter, or to hide.[18] In building, to screen usually means to mask or filter sun, wind, or just the views occupants have of one another.

A facade fills a view, enduringly, often inescapably, in embodied space. Like a face, it can express, edify, impose, or mask. In some uses, *facade* implies a false appearance to hide behind. Because a facade may bear inscriptions, whether in stone, calligraphy, fresco, flyposting, neon, or LED meshes, its full extent also becomes a frame. Traditionally, architectural composition emphasized both the facade and openings. This made the facade a two-way viewing device.

Today when the word *window* more usually means a frame for a task that appears on a screen, it is worth taking note of these related constructs of openings. Whether in architecture, graphical user interfaces, photography, cinema, or their combinations, the aperture has many vital stages in the history of information. For centuries, the architectural contexts of these apertures dominated visual culture. Then came the ascent of codex books, camera exposures, and cinematic screening as frames. For the last century, cinematic screening has dominated: the viewer is still, the frame is generally undivided, the screen lights up, and the image is seen through its moves. But now, as new display technologies flood the world, a new visual era has arrived. Metaphorically, the frame has shattered into so many fragments, each of which is itself a visual instrument. Painters have long understood how to manipulate such multiplicity of projected vision, of course. Nearly a century before stacks of virtual desktop windows, or the layering of the city with glowing rectangles, the cubists vividly fragmented space.

One core belief in media studies is that when a frame fixes a perspective, it also fixes a cultural position. The frame represents conventions of suspending disbelief. To question the frame is to expose those conventions. Postmodern scholarship learned to go beyond the visual genre to which the frame belonged, for example, into the politics or literature that influenced the frame's visual production. In today's information technology, such cross-readings and metacontexts accumulate nearby or may be summoned instantaneously. Although in any age a master such as Velázquez or Vertov could question the frame from within, digital media have generally made it easier, indeed normal, to do so from outside, in close proximity, by quick juxtapositions of ubiquitous media.

Among recent works of visual culture theory, none has looked quite so literally at the frame itself, nor so thoroughly at the frame's relations to windows and screens, as Anne Friedberg's rich history, *The Virtual Window* (2006).[19] Beginning with Renaissance architect Leon Battista Alberti's oft-cited metaphor of the window as perspectival aperture ("I draw a rectangle of whatever size I want, which I regard as an open window through which the subject to be painted is seen")[20] and closing on the multiple apertures of Microsoft Windows, Friedberg found an "age of windows" in which sitting before the frame's fixed perspective became normal, and for which cinema bore out "in retrospect, the remarkable historical dominance of the single-image, single-frame paradigm as in intransigent visual practice."[21]

The frame, whose general use arose concurrently with the use of easels, optics, and perspectival projections, comes to represent all such artifice. "The exact origins of the picture frame are somewhat indistinct, but the frame became a component element of the painting when the painting became independent from the wall," Friedberg observed.[22] A frame creates an interpretive context by setting its contents aside from their surroundings. This tends to privilege the contained, more like a treasure box than a window. A frame is an object that makes its contents an object. "The frame suggests a common position for viewing: separate from but facing it."[23]

Next comes the insight to juxtapose two separate images that invoke the same feeling, through montage, associative recall, hypermedia, framelessness; this is the core of media studies. In his influential *Language of New Media* (2002), Lev Manovich asserted the essential construct of digital media to be the composite[24]—rather than simply serving as a transparent frame, digital media now layer, substitute, and recompose.

Historical perspective on the multiperspectival paradigm is "postcinematic," an expression for which Friedberg could well be credited. The cultural distance necessary for such perspectives has quickly grown with the proliferation of virtual windows—first on the desktop graphical user interface, then with the handheld and situated gadgets of pervasive computing, now at larger-than-life scales to integrate with the built environment, and using amorphous display techniques sometimes without screens at all.

New Surfaces, Amorphous Displays

Thus the looking glass has shattered, with each of its shards becoming an instrument in itself. Each now assumes unique features and reflects specific tastes; together, the shards are experienced in a montage that is again situated in built space. Today, the proliferation of virtual windows not only moves beyond the desktop, but also beyond the static frame. Advances in display technology diversify the contexts and formats in which visual information may appear. Beside architecture appearing in projected movies, movies (and other projections) appear displayed onto architecture, as in the works of eminent artists such as Rafael Lozano-Hemmer. For more casual purposes a projector can fit in your pocket and display onto any nearby surface, such as a colleague's shirt. Sometimes fabrics themselves serve as display devices; by means of conductive thread, eTextiles already exist.

So far, the proliferation of display possibilities has run on flat screen technologies, especially liquid crystal display (LCD) technology, which are still fairly expensive and largely confined to rectangles. It wasn't really possible with the cathode ray tube (CRT) screen, which took up too much space and filled most of a desk or table top; it couldn't be hung on a wall, much less in a smaller spot like a seat back in an airplane or car. The cultural appeal of the flat screen has become so powerful that you can't give CRT screens away—many reuse centers or charity dropoff points simply refuse to take them. Thus considerable anticipation surrounds the next stage of succession, from backlit rectangular panels to new forms of display.

Electronic inks don't glow, are visible at wider angles, and don't have to be refreshed as often as backlit LCDs, which makes them much better suited for sustained reading of still text and for reading in reflected natural light. Early e-readers such as Kindle® were among the pioneers at bringing electronic inks to mass markets.[25] It has helped that e-inks consume power only to change, not to maintain, an image. Their slow refresh rate makes electronic ink displays inappropriate for rapid smooth actions, like moving a cursor or playing a video or video game, which has helped distinguish e-ink devices from more general graphical user interfaces (GUIs). Readers of literature gain peace of mind knowing that a flash clip won't jump out at them when they turn the page.

Reflective displays eliminate backlighting and yet support rapid motion. Advances in LCD technologies such as ChLCD (cholesteric LCD) display, introduced by such companies as Kent and Magink, work on scales large enough for billboards. And because reflective LCDs don't emit light, reflective LCD electronic billboards are less obtrusive. Reflective LCD display technology also works for indoor applications of arbitrary size, by means of a system of stackable 7-inch tiles. By using reflected light, wall-scaled graphics also become less obtrusive.[26] This works well on the intermediate scale of kiosks and exhibit design. It allows video (which so many people seem to want) to play without luminance, which annoys neighbors. In another variant, combining LCD and LED technologies, Transflective™ displays use either reflected or emitted light depending on conditions.[27] This provides power savings by day, when backlit display isn't necessary. It may also work well for e-books.

Relative to conventional backlit LCD, these new technologies consume less power. Thus, besides having less obtrusive display, they also have less obtrusive housings. In some cases, power consumption is low enough to run wirelessly, on small batteries or solar cells. Power is saved by replacing hungrier, earlier technologies, such as neon. In this, the most widespread opportunity is to replace incandescent city lights, such as the ubiquitous yellow of sodium fixtures, with energy-conserving LED systems. Much like the replacement of gas-burning lamps with incandescent lighting a century ago, this transition not only reduces waste but also introduces new kinds of programmable control. When combined with lighter housings and wireless power, this lets an LCD/LED display go just about anywhere.

MIT Media Lab called it "pixel liberation."[28] Whereas reflective displays escape unwanted glowing, liberated pixels escape ubiquitous rectangles. This provides prospects more in abstract data display than in conventional pictorial imaging. As William Mitchell observed in 2006, "[LED technology] breaks down the traditional distinction between computer displays and lighting systems, and provides a new and very inexpensive way of visually defining and unifying urban public spaces."[29]

In a step toward participation, many new surface technologies allow for touch. Large-scale multitouch screens let whole groups engage in visual display much as the far smaller touch screens of smartphones do individuals. The Helsinki City Wall project (2008), one of the first to apply multitouch in a sidewalk setting (figure 7.7) served as a conversational site.[30] The technology more often appears indoors and horizontally, as on exhibit tables in museums.

Multitouch floors provide the more bodily-orienting sense of engagement that comes from walking on something. As far back as the 1960s, interface pioneers such as Myron Krueger recognized this condition. Early twenty-first-century prototypes from the dawning tangible media age applied iFloor technology to information access, such as in the Alexandra Institute's 2004 project for the public library in Aarhus, Denmark.[31]

7.7 Early public multitouch surface: *City Wall*, Helsinki, Helsinki Institute for Information Technology, 2007. Photo: courtesy of Giulio Jacucci.

Although windows have been the emphasis here, note that floors and ceilings also played much richer roles in architecture before print and electronic communication. Changes of material and texture in floors communicated zones and protocols, and the design and decoration of ceilings conveyed aspirations. It is a fundamental act of orientation to look up to something. Many cultures have built this into their architecture. And it is a an impoverished culture, or at least a sign of poor architecture, where people walk around without without looking up.[32]

Embodied orientation to a surface can affect how you read and make sense of that surface. Whether an image, text, abstract data display, ornament, or just well-composed building skin, an architectural surface isn't just visual. Furthermore, the perceptions and mental constructs that result from better-balanced sensory experience produce different, and possibly richer, spatial mental models of the sites of experience.[33]

Thus it seems fair, and perhaps quite culturally significant, to value urban landscapes of information more, and the particular images applied to buildings less. Advertising supergraphics are leading the wrong way culturally. Still photos of empty buildings for glossy magazines do no better. And, of course, the notion that ornament, signage, or data systems are outside the scope of architecture now seems quite dated. Instead, the question becomes how all these visual systems aggregate.

Visual Overload Reconsidered

You might want new ways to look at this media-laden world. As the technological means for delivering superabundant stimuli keep expanding, so too must the attention skills for dealing with them.

Vision always filters; human powers of visual selection prove quite remarkable. Yet it is commonly recognized that involuntary responses exist, especially to quick movements of bright objects in the periphery. Although in the natural settings where vision evolved, those might be infrequently encountered predators or prey; instead today they are incessant, cognitively engineered, attention-seeking, artificial annoyances.

Filtering depends on more than vision. Seeing is, of course, culturally conditioned. But today that conditioning has become much more technologically mediated, for example by explosive growth in the sharing of photos, or in the perpetual use of social media as a means of filtering.

Nevertheless, filtering has its limits. You can screen out ultraviolet wavelengths with a good pair of sunglasses, but there is nothing you can wear, short of a blindfold, to screen out an electronic billboard. You can get up and walk away from a desktop computer, or put your smartphone away in your bag, but sometimes there is no way to escape an information landscape.

Thus the ambient requires some governance. As inhabitable image formats diversify, so do the occasions and ways of reading them. Much as the twentieth century amplified the ability to deliver active images within a frame, so the twenty-first is well on its way toward creating landscapes of visual production, full of communications for someone else. In these, the process of filtering has increasingly been left to the viewer. For the web, social production advocate Clay Shirky has called this "publish first, filter later."[34] But while the online world of hypertext has well-developed tools for filtering by those who think to apply them, the physical world of tagged and screen-laden cities does not.

Physical embodiment also invites foraging. Distinct from searching, where you know specifically what you are after, browsing, where you are open to whatever might appear interesting, foraging moves among clusters of stimuli. Like a bear in a mountain meadow, who moves on to the next berry patch before finishing the present one, and indeed as soon as another looks interesting, foraging in information media runs less on optimization (the best berry patch) or efficiency (getting every berry) than on perceptions of sufficiency (plenty right here—or there). Distribution in physical space brings embodiment into this kind of attention. Foraging the city involves spatial navigation, rights of access, and interpersonal social distances. It also involves formulas of visual cues matched to target populations; modern marketing combines spatial sorting with cultural filtering. Consumers avoid overload by confining themselves to familiar "brandscapes."[35] Nevertheless the least predictable zones of the city are often the most interesting ones. Besides even in the most formulaic settings, advertisers cannot assume that anyone is looking.

Thus, as a way to imagine the degree of change wrought by the ascent of filtering, consider the oxymoron "unnoticed spectacle." Much of mid-twentieth-century critical theory assumed that, with radio and especially television being run by just a few centralized networks, most people would notice a single dominant feed and that "spectacle" would reflect how much of that feed was engineered as distraction, presumably for political purposes. As voiced best by Guy Debord, spectacle not only commanded view but also furnished the terms of viewing: by providing the times, places, language, talking points, and subject matter for public attention, it was the perfect frame.

Today, so much has been made into spectacle that little of it amazes. So many channels exist that every group of viewers has one just to its liking. So many visual media exist that no single medium, not even television, nor electronic billboards, can claim command of its viewers. Recording, fast-forwarding, sharing, linking, exposing, rating, bombing, sousveilling, and so many other actions of many-to-many communication reduce the chances for any one feed to be noticed. Although many people remain passive captives of media monocultures, none of the former one-to-many channel owners can so confidently claim, as their twentieth-century counterparts did, that "we" think this particular way or that. Even the most watched event of the year, whatever that may be for a particular subculture, is watched alongside several others, intermittently. The contents of each frame, engineered ever more crassly to capture attention, nevertheless fail to capture attention, which instead drifts and marvels at how, wherever you look, so many frames blink and glow.

With so many new relations among windows, screens, frames, and facades now filling everyday space, watching has become less important, and foraging has become more so. Having more options improves the chances that you will find enough interest somewhere. Often you can discover that without the concentrated effort of dedicated search. You work at a lower level of detail, and a higher sense of drift. If there seems to be enough here, you might be open to more of it, and scan it more closely, or with more particular intent. You take cues from surroundings about what to be looking for. Context and sensibility intertwine.

The ambient is such a relationship. The return of inscriptions to building scale recalls a time before framed apertures. To inhabit a patchwork continuum of glowing surfaces is to look less at any one. To move among so many screens as an environment in itself is to stop less often before any one of them. To enjoy their multiplicity as an embellishment of their place is to look less through any one of them as if a threshold to somewhere else. In the words of Stephenson: "Despite their efforts to stand out, they are all smeared together."

7.	FRAMES AND FACADES
Main idea:	Explosive growth in number, formats, and contexts of situated images
Counterargument:	Corporate media and spectacle
Key terms:	Frame, glow, amorphous, postcinematic
What has changed:	People move around with and among displays
Catalyst:	Many new display technologies
Related field:	Visual culture theory
Open debate:	Shared visual experience?

Architectural Atmospheres 8

The ambient began as embrace by air, or as the ancients expressed it, "aer ambiens." Now many fields have rediscovered this embrace. Air itself is "in the air." So besides all those tags and screens as objects of attention, consider the ambient as something you occupy, with fully embodied cognition. The built world reshapes attention sensibilities not only through its configuration but also through atmospheric comfort. Architecture encloses air, conditions it, and, in the process, conditions attitudes too.

Composing Air

As many inquiries into air now do, start with oxygen. Throughout the last decade, many thinkers have recited stories of Joseph Priestley's discovery (in the late eighteenth century) that air is not a single pure element, but a composition that includes oxygen.[1] In the late twentieth century, "Oxygen" became the name of an early MIT project on computing "as pervasive and free as air."[2] Later that idea was taken up by the European Union research agenda on ambient intelligence.[3]

Air has always had its metaphors. Traditional myths and legends held that fates were the exhalations of the gods. Illness was believed to come from ill winds and nocturnal miasmas. Before knowledge of oxygen, combustion was thought to occur across a metaphysical ether called "phlogiston." Air had always been fouled by fires; anyone could see as much. The modern change was to make science and policy about it. Technological historians take interest when several fields simultaneously arrive at workable concern. Air was one of these, two hundred years ago; and today it has become so again in new ways. Through technology, humanity now engineers air, both at building and planetary scale. Willis Carrier, the pioneer of air-conditioning, called the former "manufactured weather." Today, at the latter scale, that expression has staggeringly different implications.

The philosopher Peter Sloterdijk traced the fear of technologically modified air to the poison gas attacks of World War I. In sociologist Bruno Latour's oft-cited paraphrase of Sloterdijk: "You are on life support, it's fragile, it's technical, it's public, it's political, it could break down—it is breaking down—it's being fixed, you are not too confident of those who fix it."[4]

Contrary to centuries of beliefs in the West, then, it turns out the invisible stuff between visible things is of consequence. The upkeep of air is everyone's concern. Of course not everyone realizes this. Why else would people sit in their cars at an overlook point, with engines idling and windows rolled up on a perfect California day? But now, as planetary atmosphere becomes something for everyone to think about, the rediscovery of air has become a prominent cultural theme.

Atmospheric Arts

In a parable for reconnection between the planet and the arts, a 2008 retrospective on British romantic landscape painter J. M. W. Turner, thought to be the first to have written, "Atmosphere is my style,"[5] called attention to luminous air as subject matter. That the skies were full of volcanic ash in 1816 made the sunsets Turner dramatized even more dramatic. Though just an interesting circumstance for the artist, this event held far greater significance for viewers in 2008.

Air can be subject matter, expression can be atmospheric, and sometimes these two can converge. Many familiar examples exist in film, whose reliance on montage seems fundamentally atmospheric, and whose capacity for fast action cuts is so often abused that artistic restraint seems all the more memorable. Take Sergio Leone's *Once Upon a Time in the West* (1968), for example. In an opening scene that lasts several long minutes, two gunmen sit motionless in the shade of a railroad station, with only a creaking windmill and buzzing flies to be heard, as they wait for a train to arrive from great distance through the shimmering desert air.[6]

Whether you call it "atmospheric," "ambient," or "minimalist," work that refrains from jumpy, noisy, foreground effects seems ever more the exception, and an ever more necessary haven, in an age of attention-seeking overstimulation. Although this inquiry is not the place to unpack a critical aesthetic of the ambient arts, it would not be happening without those arts. Embedded sensors, actuators, and computation have enabled a new aesthetic genre in environmental monitoring (figure 8.1). The ambient arts have several ideas to offer on perception that are worth considering here. These may help show why to

8.1 Ambient sculptural display of outdoor air movement: Jason Bruges, Anemograph, Sheffield, 2006. Photo: Jason Bruges Studio.

approach the question of attention from the perspective of architecture.

As noted in the chapter on embodiment, Gernot Böhme recast aesthetics into a general theory of embodied perception, where "the primary 'object' of perception is atmospheres."[7] Böhme asserted a basic postmodernist thesis: by reaching into all aspects of life, a new aesthetic departs from high art based on cultivated taste. "The primary task of aesthetics is no longer to determine what art is, and to provide means for art criticism, [but to describe] the full range of aesthetic work, which is defined generally as the production of atmospheres."[8] Note how this cultural diffusion coincided with the scientific admission of "preontological" cognition, which, in turn, led to the current emphasis on "affect," that is, communication without overt messages. Böhme thus traced the lineage of affect to Walter Benjamin's famous expression of "aura."[9] Whereas aura was associated only with high art objects, however, affect is anywhere, and aggregate.

Twentieth-century aesthetics defined an *atmosphere* as an "overall mood produced by individually indiscernible moves in a medium." Thus an atmosphere may arise when many different things in a room share the incidental property of color. Whether in painting, film, or music, the twentieth-century arts generally moved away from traditional themes and icons to concentrate on the intrinsic physical traits of their particular mediums. For example, when Claude Debussy put aside conventions of classical exposition for an emphasis on the timbres of discrete sound events, it was considered atmospheric, and invited a new kind of attention. For this, Debussy is often credited as a founder of both ambient music and some of the first modern music.[10]

The high modernists of the mid-twentieth century stripped away traditional symbolism, reducing art to its actual surfaces. Visual theorists such as György Kepes and Rudolf Arnheim addressed cognitive mechanisms directly, through the principles of gestalt psychology as they interpreted them, and visual artists often did so, too. This was especially evident in painting, where nonrepresentational emphasis on the paint itself pushed the medium toward its most atmospheric possibilities, such as the floating color fields of Mark Rothko. In music, it led to pattern-and-process pieces, such as those of Terry Riley. Many such works came to be known as "minimalist." Although this label has been inappropriately applied to many phenomenological artists who didn't regard themselves as minimalist, it makes a point. Art has to do with the unnamed. It operates without ontology or semantics. This role became clear in the 1960s, for for example when artists challenged the boundary between sculpture and painting, forcing viewers into an awareness of the space of the gallery itself. The sculptor Robert Morris, often credited with laying the theoretical groundwork of minimalism, once defined the mission of art as "maximum resistance to perceptual separation," a phrase that prefigures later formulations of atmospheric aesthetics.[11] But fellow sculptor Donald Judd may be the artist most often mentioned when people are asked to name a minimalist today. To visit his installations in Marfa, Texas, remains an act of high fashion.

In her widely read book on the relations of art and cognitive neuroscience, historian and critic Barbara Maria Stafford has explained the visual arts' move beyond illusory tricks (e.g., trompe l'oeil or anamorphism) to a richer exploration of how the brain constructs perceptions of environmental coherence. For, as embodied knowledge theorists have come to explain, the brain makes use of objects and surroundings to construct, master, and maintain perceptual schemas that do not rise to the level of processed symbols and signs, but are instead experienced in lower neural pathways. Stafford has investigated the role of images in these constructs and the power of some patterns to create, not merely depict, holistic environmental perceptions. She has explained how phenomenological coherence takes the luminous works of artists such as Dan Flavin and James Turrell beyond optical deception into "unfurling metaphysical experience," or at the cognitive level, "neural holism."[12]

Atmosphere, then, goes past the play of signs, which so overloads everyone, to a more intrinsic kind of information. It does so through embodied cognition, which works well at the scale of built volumes.

Architectural Atmosphering

As this inquiry has explored with respect to attention, embodiment in architecture and the city plays important cognitive roles. Although contextual sensibilities may arise from natural, mythological, or networked technological configurations, or of course from personal and social affinities, the built environment also has influence. As noted in the chapter on fixity, architecture provides mise-en-scène for communications, configures the habitual play of expertise, and can help to restore attentional capacities. Yet the most basic role of building is to provide atmospheric comfort. A rediscovery of air thus invites quite a different way to think about the relationship of architecture and attention.

For much of the last decade, the discipline of architecture has turned away from attention-seeking novelties in digitally generated form toward more slowly knowable atmospheres in space. Besides being a reaction to the rediscovery of air, this was also a reaction to the kinds of perceptual overload brought on by the rise of computers. Perhaps most welcome of all, a new interest in atmosphere was a reaction against the dulling uniformity of corporate buildings, where the only interesting place to look was your screen.

Surely the most emblematic work of that decade was by the Danish phenomenological artist Olafur Eliasson, who is best known for The Weather Project installation (figure 8.2), at London's Tate Modern, which by some estimates had two million visitors over its six-month run (2003–04). For many, The Weather Project signaled the emergence of designers willing to explore new topics in environmental installations, small climate objects, and physically atmospheric works.[13] Eliasson soon

8.2 Olafur Eliasson, The Weather Project, London, 2003–04.
Photo: Thomas Pintarec/Wikimedia Commons.

became one of the most sought-after installation artists alive.[14] In her 2006 profile of him, *New Yorker* critic Cynthia Zarin explained how Eliasson "was deeply affected by the work of the phenomenologist philosophers, especially Edmund Husserl—with their emphasis on the individual experience of reality—and by Lawrence Weschler's biography of Robert Irwin, *Seeing Is Forgetting the Name of the Thing One Sees*."[15] (As noted in the chapter on the ambient, Husserl's phenomenology lay at the foundations of embodiment.)[16] The Weather Project required the participation of the viewers for completion. "Rather than minimize [the great height of the hall], Eliasson doubled it" visually,[17] with mirrors on the ceiling; he made the space outdoor-like with a giant disk of a sun and made it slow with a long cycle of clouds from mist machines. Viewers would lie in this space, sometimes for hours, sometimes losing perceptual separation between themselves, the setting, and the art, and sometimes even forgetting to go on to the picture galleries.[18] Participants remember the project atmospherically, which may be why it can be cited so often without becoming a cliché. Recall Debussy's conception of atmosphere, where the effect of the work at any given moment can't be traced to any particular component. In The Weather Project, viewers couldn't even trace the effect externally, outside their participatory nonfocal attention.[19]

In other words, design practice can literally make sense by creating phenomena that help construct signless mental representations. In an age where disembodied stimuli otherwise command an ever larger share of attention, atmospheric design can restore some cognitive balance.

Yet, for at least half a century, air has usually been designed to go unnoticed. A fully automatic environment is more efficient;

it means one fewer thing to think about. But now that creed gets called into question. Uniformity no longer seems attainable, affordable, healthy, or technologically necessary. Today's sensor and actuator networks allow finer adaptations than were practical before.

Many recent works in architectural atmosphere emphasize these engineering prospects in themselves, without need for staging in grand spaces like the Tate Modern. Digital production techniques allow unprecedentedly precise control over the composition of minimally incremented differences. In one of the most provocative bodies of work in "atmosphering," the Swiss architect Philippe Rahm rose to prominence simply by reintroducing light, heat, and ventilation as objects of design. Rahm's work makes environmental contrasts figurative. He speaks of "thermogenesis."[20] For instance, a project for a house creates a gradient from very warm to very cool, instead of uniformity. Rahm's works engineer physiological rather than visual response; they dramatize embodied cognition. Perhaps best known of these is the Hormonorium, or "melatonin room." An installation at the 2002 Venice Architecture Biennale that simulated the perceptual shift you would feel in an alpine snowfield, this room "established a continuity between architecture and human metabolism, between space, light, and the endocrine and neurological systems" by means of anomalies in air and luminance. Nitrogen was pumped into the room so that oxygen levels felt like those at high altitudes. The floor emitted very bright daylight spectrum illumination from below, like the albedo of snow, as light from below is less naturally screened by the body than light from above. Response to this intensity decreased secretion of melatonin, a hormone that helps regulate wakefulness and sometimes libido, even as the rarefied oxygen increased secretion of endorphins.[21]

Meanwhile, in *Atmospheres*, perhaps that decade's most-read text on atmosphere in architecture, Swiss architect Peter Zumthor described "embodiment, material presence, the sound of a space, the transfer of warmth, incidental surrounding objects, drifting between composure and seduction, concentration in enclosure, proximity and distance, bringing light into shadow."[22] "Calming," the principle so essential to Mark Weiser's vision of ubiquitous computing, also has a place in Zumthor's architecture of material presence. "There are practical situations where it is more sensible and far cleverer to introduce a calming effect, to introduce a certain composure . . . Where nothing is trying to coax you away, where you can simply be."[23] Although Zumthor fits many stereotypes of the lone poet, and so gets labeled as one of the romantics that modernists like to dismiss, his interpretations remain influential. When, in 2009, Zumthor received architecture's highest honor, the Pritzker Prize, for many this legitimized atmosphere.

Primal relations connect air with warmth, and thus also with fire and pollution. In *Fire and Memory*, Luis Fernández-Galiano interpreted buildings mainly as channels for energy, contrasting an "exosomatic" kind of energy that surrounds for warmth with the more usual kinds of energy that power production. In what amounts to a thermal theory of intrinsic information for warmth, memory of surrounding energy flows gives a structure both to construction and perception; the form of space recalls the flows of energy.[24] By abandoning many traditional forms, modern building has caused an amnesia, or at least a dulling of "the eye and the skin."[25] Given how much of this amnesia comes from humanity having moved indoors for the last century, it is now fashionable to say that this loss has been "conditioned by air."[26]

Conditioned by Air

Few technologies have been so expressive of modernity on the whole, and yet so inexpressive in themselves, as air-conditioning. Here is an extreme case of technology so successful that it disappears into everyday life. Indeed, many of the places developed since air-conditioning came into everyday use in the mid-twentieth century would be uncomfortable or even uninhabitable without it.[27]

Air-conditioning has been an attentional technology from the start, with the purpose of forgetting the climate and accomplishing more work. Willis Carrier foresaw a brave new world where "every day would be a good day."[28] Not only the temperature of air but also its humidity and purity were to be conditioned. There was design appeal in the fact that one airstream handled all of these problems.[29] But there was appeal as well in command and control itself, especially of productivity. Le Corbusier, architecture prophet of the hermetically sealed building, foresaw one temperature to fit all, 18°C (64°F).[30]

Like many technologies on which humanity has become dependent, air-conditioning began as a luxury. Even into the 1960s, most cars still lacked it. Specialized workplaces were the early adopters, first in manufacturing operations and then in offices, where the consistency of output depended on environmental conditions. In sultry Washington, the House and Senate chambers got air-conditioning in 1928, the White House in 1929, the new Supreme Court building in 1932, and the remainder of the capitol complex in 1935. Movie theaters and department stores first brought air-conditioning into public use in the 1920s and 1930s, and they made an attraction of it. The atmosphere was a goal in itself; on some summer evenings, it became the main reason to go to the movies.

Meanwhile, architects experimented with the technology. The Philadelphia Savings Fund Society building (1932) was the first fully air-conditioned office tower on the East Coast, the first to integrate technology in ceilings, and among the first to display its huge text logo (PSFS) on the skyline.[31]

The systems integration of the climate-controlled glass tower of the post–World War II era soon fueled fantasies, epitomized in the television cartoon sitcom *The Jetsons* (1962–63), and immortalized for architects in the 1964 exhibit Living Cities, by the London-based the futurist group Archigram. With unlimited actuation and boundless cheap energy, comprehensive mechanical systems became not just fittings to preconceived structures but buildings in themselves. Although the age lacked embedded computation, these futuristic visions of a fully responsive architecture were nevertheless computationally inspired, according to the cybernetic agendas (and fantasies) of the time.

To open *The Architecture of the Well-tempered Environment* (1969, and an enduring a classic in this field), architectural technology historian Reyner Banham apologized for architecture's neglecting the "whole of the technological art of creating habitable environments." For, "in a world more humanely disposed, and more conscious of where the prime human responsibilities of architects lie, the chapters that follow would need no apology, and probably would never need to be written."[32] That said, Banham saw air-conditioning as the key to full architectural expression:

> Firstly, by providing almost total control of the atmospheric variables of temperature, humidity, and purity, it has demolished almost all of the environmental constraints on design that have survived the other great breakthrough, electric lighting. For anyone who is willing to foot the consequent bill for power consumed, it is now possible to live in almost any type or form of house one likes to name in any region of the world that takes the fancy.[33]

Charting the history of steam heat, sanitation, elevators, huge buildings, and technology outside the workplace, developments catalyzed by that most truly fundamental agent of modernity, electrification (without which there would be no air-conditioning), Banham found the origins of a machine aesthetic in the early twentieth century. Much as electricity moved work away from the furnace, and light and materials to the work site, so pervasive computing is moving work away from the office and desktop, and into the very sites to which it applies. Much as electric motors were embedded invisibly into things not considered "electric motor applications," such as the refrigerator, so computer processing and memory are now embedded into things not considered computer applications, such as vehicles. And much as electrification gave rise to building environmental control technologies, so pervasive computing is giving rise to adaptable, socially networked versions of those technologies. Smart green buildings do more with less. They promise escape from dulling atmospheric uniformity.

In *Thermal Delight* (1979), another classic text from the era when architects began to question the sealed glass box, and one that still captivates nonarchitects curious about contextual attention, Lisa Heschong explored adaptation to climate, whether through migrations, protective furs, shells, and skins, or the building of structures. This is natural. Moreover it is intrinsic. "When our thermal senses tell us an object is cold, that object is already making us colder. If, on the other hand, I look at a red object it won't make me redder, nor will touching a bumpy object make me bumpy." The sense of touch, she explained, recognizes heat flow and not static temperature: "If I touch a piece of metal and a piece of wood that are both at room temperature, the metal will feel colder because it absorbs the heat from my hand more quickly." There is delight in these differences. By contrast, Heschong likened the "standardized comfort zone conditions" of climate-controlled buildings to the prospect of nourishment by pills, injections, or "astronaut's nutritional goop." For, even though "eating is a basic physiological necessity, no one would overlook the fact that it also plays a profound role in the cultural life of a people."[34] So does thermal delight. Many adaptations to different thermal conditions, such as different spaces for different times of year, layers of clothing, and, of course, human-operated layers of buildings themselves involve the pleasure of using them. It feels good to put on a sweater, to open a window, to move in and out of the shade as you walk down a street.

The rise of the ambient invites new integrations of high and low tech, and of thermal with other delights, even as the rise of energy costs makes uniformity far less affordable anyway. Indeed, advances in building technology and in the understanding of air currents themselves make the quest for uniformity seem naive scientifically, compared to what designers can now accomplish with fluid mechanics, boundary layers, and passive cooling systems.

"Good-bye, Willis Carrier," wrote architectural technology critic Michelle Addington. "Hello to we know not whom."[35] But, in this age of smarter materials, surfaces, and buildings, and of ever costlier energy, that might be hello to personal environmental management, shared sensor fields, and ever higher levels of actuation. And good-bye to the project of total technological control, disregard for timeless low-tech solutions, and amnesia for the ambient delights of the unmediated world. Good-bye, in short, to what renegade novelist Henry Miller, writing fairly early in the history of the one technology that most epitomized his gripe with America, called the "Air-Conditioned Nightmare."[36]

Filling the Air

Alas, for the moment, the nightmare worsens, for there is one more technology of excess uniformity to consider. No environmental history of information, nor any reconsideration of attention, would be complete without a look at that most ubiquitous, unrelenting, inescapable atmospheric scheme of all: background music.

What ubiquitous information technology does for atmospheres in architecture is to create a nearly universal impulse to fill them. Much as screens have filled almost every empty spot in shared built spaces (just try finding a bar without television), so the more purely atmospheric technology of background music has filled almost every empty volume (just try to find a store, shared workplace, or restroom without audio). Miniature personal players and earpods let you bring music almost anywhere. And, to avoid that awkward few seconds of dead air you get on a dive into a swimming pool, you can now install underwater audio speakers.

"I bring earplugs on planes to block out . . . the insipid pabulum of the boarding music," *Travel & Leisure* senior editor Peter Jon Lindberg ranted.[37] In the ubiquitous use of audio streams as branding devices, the chief offenders are behavioral marketers—"puppeteers who don't play music so much as deploy it." Although Lindberg's beat is high-end hotels and spas, where "new age" music is pumped through those underwater speakers, the audio perfume goes just about anywhere else. "A friend of mine went in for an MRI," Lindberg recounted, "and had to endure not only her own claustrophobia but also the clinic's cheesy piped-in sound track—45 minutes of continuous soft hits to the head." The programmers' assumption that nobody is listening is exactly false; at any given moment, somebody is totally stressing out.[38]

Muzak® achieved mainstream acceptance about the same time that air-conditioning did. The name itself is a conflation of "music" and "Kodak," presumably to imply push-button ease of use. The service began under the name "Wired Radio" in 1922, took on its current brand name in 1934, and played a role at stress reduction for U.S. troops in World War II. It then steadily ascended toward its heyday of the 1960s, where in now-legendary sessions, professional ensembles such as Hollyridge Strings committed to vinyl their defamiliarizing orchestral renditions of rock standards like "Strawberry Fields Forever."[39] Muzak® streamed over leased lines to tens of millions of office workers, hotel guests, and yes, elevator passengers. In one of the more savvy social histories of any ambient technology, Joseph Lanza explained that "background music companies never considered elevators their biggest market, however. Their primary customers were places of work and recreation that used music as a mood boost, just as radio had orchestrated private homes."[40]

Lanza traced audio behavioralism to Edward Bellamy's 1887 utopian science fiction novel, *Looking Backward*, cited in a Muzak® brochure on the company's fiftieth anniversary. Musical mood boosting also appeared in Aldous Huxley's 1932 science fiction classic, *Brave New World*, where "Synthetic Music" made with "hyper-violin, super-cello, and oboe-surrogate" contributed to the "agreeable languor" of a totalitarian future.[41] Lanza found the surrealism of Muzak® to be its most significant trait. Muzak® orchestrations took compositions into hyperrealities (first on Lanza's picks of twenty "metarock" essentials is "California Dreamin'," as interpreted by Hugo Winterhalter).[42] Also surreal was Muzak®'s delivery from disembodied sources: filtering out of ceilings and walls, back when that was still considered odd. But, above all, Lanza found mood music surrealistic with attention: "For a more clinical definition: mood music shifts music from figure to ground, to encourage peripheral hearing. Psychoanalysts might say that it displaces attention from music's manifest content to its more surreal latent content."[43]

Of course, not everyone thinks that silence brings only discomfort and boredom, even when waiting. Critics of mood music, some of them vociferous rockers even, do value sustained silence from time to time. Indeed, you might ask whether someone who needs to fill every moment with music, movies, or messaging isn't somehow existentially challenged.

Critically acclaimed ambient music does exist. Some ambient pieces, such as those written or inspired by Karlheinz Stockhausen, come from the traditional disciplines of Western music. Others seem more countercultural, such as the sounds that stream over "Drone Zone" Internet radio.[44] For with the advanced sequencing capacity and timbre inventions of the latest software, music needn't have a tune as such to be legitimate. Yet much as people have been conditioned by air to have a much narrower thermal comfort range, so they may also have been conditioned by background music to have a much narrower audio comfort range, a range whose very narrowness and predictability are comforts in themselves. Lindberg could confidently predict hearing Sade croon "Smooth Operator" once again in a frequent flier lounge.[45] Anything more experimental would be obtrusive. Probing "elevator noir," Lanza observed that even Brian Eno's inexhaustible standard, *Ambient 1: Music for Airports* (1977), wouldn't play in Pittsburgh. Indeed, when they actually played *Music for Airports* in the airport there, throughout the concourses for ten hours a day, across the nine days of the 1982 Three Rivers Arts Festival, "many patrons reportedly complained to airport personnel about this uneasy listening and asked that the regular background music be restored."[46]

In early 2009, the Muzak® corporation filed for Chapter 11 bankruptcy. At the moment the industry leader appears to be DMX®, whose streams may reach as many as 100 million listeners, with or without a chance to click "Accept." Playlist matching to lifestyle category has come a long way since elevator music. The sensory effect has become a carefully formulated component of induced consumption. Thus, from the "About Us" page at dmx.com:

> DMX—where great brands make sense. In a world where the average person sees 3,000 advertisements a day, the only way to stand out is by delivering a multi-sensory experience customers won't forget. An international leader in multi-sensory branding, DMX has been creating unforgettable

> brand experiences for commercial environments since 1971. The first music service to license and program original artist music, DMX has rigorously researched and tested the effects of music, video, messaging and scent on human behavior. By integrating them into a single compelling experience, we help clients drive repeat business and build brand loyalty.

If you have a problem with this, you can always just bring your own headphones. So far, nobody has engineered noise cancellation on any larger scale.

Reengineering for Air

Consider a move away from conditioning and filling the air, toward appreciating air for itself, and including atmosphere in a better balance of attention. Just what extent of technology do you want? For being here now, there is nothing quite like taking a moment to open or close a window yourself. And for the better building performance necessary to an age of costly energy, new kinds of monitoring, feedback, load balancing, and small-scale tuning now arise. Smart green building could be either much more automatic or much more participatory, however. Much as twentieth-century uniformity proved neither affordable nor physiologically healthy, so twenty-first-century responsiveness has its limits as well: total adaptability might be neither affordable nor psychologically healthy. Architecture's fixity, reliability, and persistence count for plenty, especially in the workings of attention.

Augmentation by embedded computation makes buildings far more adaptable, and this pays. Although theaters, restaurants, and other such sites of social gathering already engineer their atmospheres quite carefully, as does the home, it is in the workplace that new building technologies count for most. Energy conservation and greater worker efficiency have made investment in better workplace technologies far more financially feasible than it was in the heyday of the sealed glass box. The workplace tends to be the most wasteful of energy, where people most often put up with technologies they detest, need to dress in extra layers to counter the air-conditioning, wear headphones to escape the background music, or step outside to get a breath of fresh air.

Increasingly, the use of sensors, actuators, and networks in buildings makes their atmospheric control less rigid, and their conditioned air more breathable. Today's building skins have come a long way beyond the glass curtain wall. With better feedback and kinematics, operable windows are a possibility again. Alas, this requires adaptations. Much as people protest when made to park more than a block from their destinations, likewise enough have been so conditioned by air that any inconvenience meets with protest. Thus, even though public health studies clearly indicate the benefits of taking the stairs when going just one flight up or down, implementing skip-stop elevators to encourage this practice is usually met with outcries of indignation.

Such outcries greeted the recently constructed Federal Building in San Francisco, the first U.S. high-rise in decades to be built without central air-conditioning. This building's "living skin" monitors atmospheric conditions, creates passive chimney effects, and triggers active vent motors.[47] Evidently, the building's freedom to change conditions was not matched with the workers' freedom to change their work location or orientation with them. And even though millions of dollars were saved in construction and energy costs, now someone's papers are getting blown around.[48] It doesn't help that signature architects have gone for decades without bringing atmospheric performance into their designs. But priorities have changed very quickly of late, and sometimes signature designers (Morphosis, in this case) are just the ones able to push owners and builders to try something different. Novelty is now less about achieving attention-seeking sculptural form and more about achieving attention-restoring atmospheres.

Still, the technology does advance. Other projects have achieved passive airflow systems in far less forgiving climates than San Francisco's, often through the use of double skins and sensate elements. Thus the Manitoba Hydro Tower in Winnipeg (figure 8.3) runs mainly on a vertical stack effect and employs geothermal heating and cooling. For the few weeks each year when the northern prairie climate is actually pleasant, the tower has windows that open.

To futurists, the building skin becomes the main locus of twenty-first-century advances in construction, as designers pursue the long-running fantasy of a fully cybernetic architecture, something well beyond the visions of Archigram and the Jetsons. Thus, in biomimetic buildings, tensile structures morph incrementally at the joints of tiled surface membranes with shape memory alloys as actuators, using a time-averaged compilation of proximate motion data as input. Like a flower opening, the surface of such structures changes form quickly enough that it appears perceptibly different each time you glance away and back at it, but slowly enough slowly to appear static under your constant gaze. This makes it atmospheric. Whether there is any practical advantage in such kinetics seems open to debate. That said, a fully responsive enclosure seems inherently practical to its enthusiasts. Walls move about to resize spaces to the needs of the

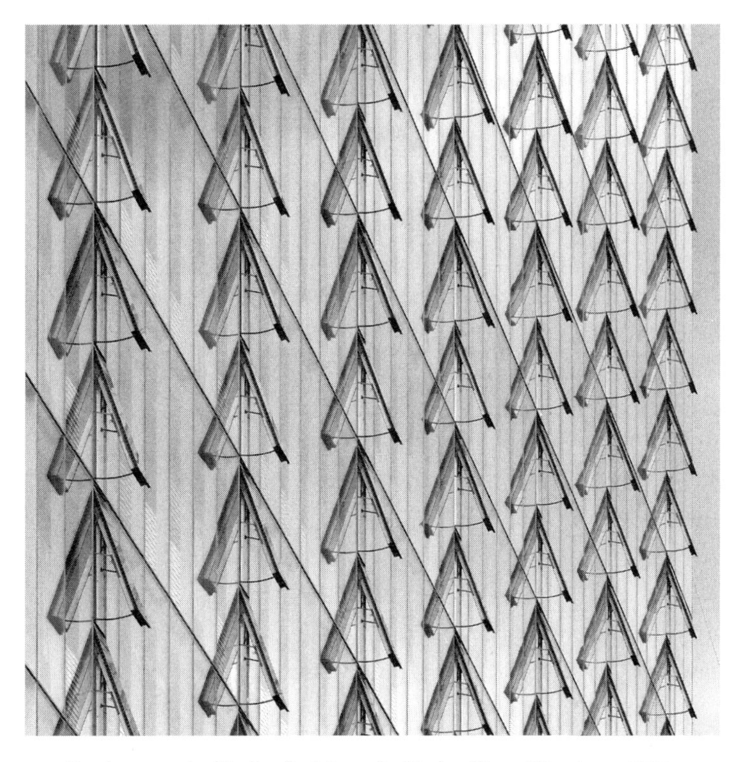

8.3 Biodynamic double facade: Manitoba Hydro Place, Winnipeg, 2008 (architects: KPMB). Photo: Tom Arban.

moment. Openings expand and contract with the weather. Building envelopes filter and breathe on demand. Data murals suggest the general co-presence of colleagues or display particular long-term derivatives of organizational performance. Fixtures interoperate, and remote-controlled devices from sunscreens to security doors to the coffeemaker serve as "proprioceptors," in the parlance of cybernetics. Although some architectural components of these fantasies are in fact real, the category error is to imagine that all are so. To expect the ultimate role of an information layer to be reconfiguring form itself belies architects' emphasis on form. How stasis could be newly appreciated, and made more usable, in an age of electronic flux gets less speculation.

The stillness of buildings does have advantages, especially with regard to attention. Embodied attention depends on the fixity and persistence of enough phenomena to provide cognitive building blocks for more expert cognition in action. From what is now known about activity theory and attention, a totally fluid and endlessly reconfigurable environment would be counterproductive. Inhabitation needs fixity, like a river needs banks. Total stasis may be dulling, but complete flux would be stupefying.

Cameras, motion sensors, noise level monitors, air quality measurements, and a variety of other instruments today make almost any new building a smart building. Embedded processing and memory bring averages, ranges, and patterns, or sometimes even complex adaptive learning, into the process of environmental management. A neural metaphor applies. Microsoft, for one, has long referred to "digital nervous systems." Unfortunately, passively living within sensate environments too often produces as much psychological unease as physiological comfort. Security systems, the most ubiquitous, visual, and passive of sensate building layers, likewise breed distrust at some higher level, even as they ensure basic personal safety. Furthermore, like background music, most of the layers give you no chance to opt out or even to tweak the resulting conditions. Your mere presence may imply consent. As in music, so in warmth, lighting levels, airflow, and just about any atmospheric condition, nothing is going to please everyone. Furthermore, despite the unquestionable benefits of today's more sensitive feedback control systems, no managed environmental condition is going to be fully stable, predictable, or uniform. Because building engineers know that occupants compensate too much, too late, a local aberration such as a drafty spot may not be subject to "end user control." Occupants, for their part, feel frustrated when denied participation despite frequent use. You might say that, as more buildings get nervous systems, more occupants get nervous.

Cooperation can both reduce these anxieties and improve physical conditions. This begins with the building becoming a social software interface in itself. According to Norbert Streitz, who coined the expression "cooperative building" in the 1990s, "with the disappearing computer, the 'world around us' is the interface to information and for the cooperation of people. In this approach, the computer as a device disappears and is almost 'invisible.'"[49] In other words, social sharing has been important to the ambient information agenda since long before Twitter. An appropriately embodied interface provides awareness not only of co-presence and gatherings, but also of building performance itself. This gives occupants a stake in seeking environmental quality, and sometimes also in managing it. In a recent survey of some 6,000 office workers across sixteen U.S. cities, IBM found nearly a two-to-one majority "willing to help redesign their workplace to make it more environmentally responsible," although only about a quarter said their buildings "adjust environment automatically based on occupancy."[50]

Despite rapid advances in smart green building, few projects demonstrate personalized or social interfaces to comfort management. Again, few engineers find that desirable. Pervasive computing

8.4 Persuasive design in building performance gauges: Building Dashboard® Kiosk, Lucid Design Studios, 2011.

does change this problem, however. Wireless sensor technologies and embedded memory systems do give higher resolution, finer sensitivity, and greater stability to building technology feedback control systems. Participatory monitoring appears on handheld apps and at office kiosks (figure 8.4). These systemic properties allow greater local variation, and greater participation in control.[51] Thus where the chaotic inconsistency of so many user inputs would have unbalanced an environmental system in the past, and been considered counterproductive to goals of efficient uniformity, today it might not. Consider the rise of human computer interface design. Back when computers were expensive, software was designed to conserve machine cycles, and to consume those in the name of usability was considered wrong. Today when machine cycles are nearly free, attention has become the scarcest resource, and both usability and aesthetics have become foremost software design considerations.

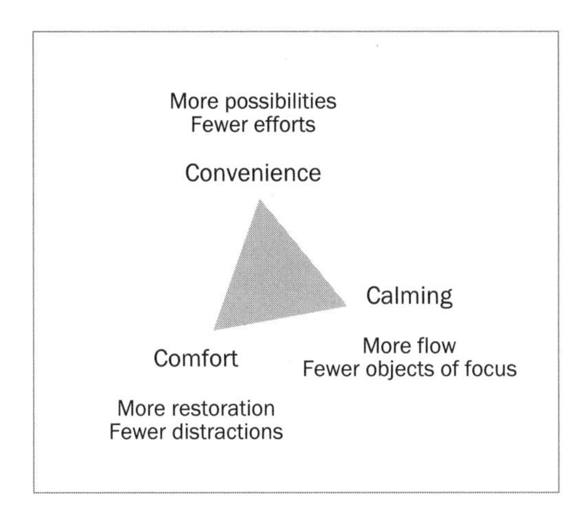

8.5 Comfort, convenience, and calming.

The technology now exists to let you override automated systems, identify the right controls to tweak to anticipate recurring conditions, and not just to overcompensate after the fact. Architecture technology researcher Raymond Cole has emphasized the need to reconcile human and technological inputs in smart buildings. Cole compares technological, organizational, and human intelligence and asserts that, for better or worse, all buildings teach or neglect some kind of social pattern of environmental responsibility. Thus, "increasingly, research confirms the importance of having some level of direct control over the environmental conditions in the workplace as being paramount to occupant satisfaction."[52] Yet it is important not to conflate comfort, convenience, and calming (figure 8.5). All must figure in what soft cities pioneer Dan Hill has called "the personal well-tempered environment."[53]

For the technology of atmosphere, and with respect to attention, the idea of cooperative buildings raises two important questions. Does the psychological comfort of having a stake in environmental management offset the physiological discomfort that may result from a tendency to overcompensate or from how no atmospheric state pleases everyone? And does participation in environment provide some grounding or calming, and thus a balance in kinds of attention, or does it become just one more annoyance adding to a sense of overload?

8.	ARCHITECTURAL ATMOSPHERES
Main idea:	Ambient is not uniform; atmosphere is design subject matter.
Counterargument:	Air should go unnoticed
Key terms:	Atmosphere, conditioned by air
What has changed:	Smarter, greener building
Catalyst:	Environmental awareness
Related field:	Architecture
Open debate:	Do inhabitants want to participate?

Megacity Resources 9

Urban computing now comes of age. Beyond the responsive room, the outdoor screen, or the location tag, information media permeate the built environment and form urban resource networks. At this scale, it becomes harder to study attention itself, yet easier to understand inhabitation and usability as one.[1]

As mobile technology remakes attention at street level, it meets the embedded. Everyday transactions use and create long trails of data. New layers remake experiences of transit, shopping, basic utility connections, and, in some cases, even the allocation of housing. Although the usual image of smart cities has been wealthy and utopian, the more profound significance of their new information layers may lie in new living patterns across the multicentered urban archipelagoes sometimes known as "megacities."[2] Active participation in emergent networks helps make local habits and routines comprehensible. Wherever people improvise organizations to get on with life amid the chaos of new settlement patterns, ambient information plays a part.

Many of these provisional arrangements provide access to infrastructure, identity to community, or opportunity for local business. Unlike global finance, which tends to operate in disembodied and disengaged ways, these ad hoc arrangements operate on the ground, in small transactions that can seldom be predictably formulated—maintaining, and sometimes even increasing, the kinds of human, social, or natural capital that remote corporations may not even recognize. As such, they demand new approaches of design, research, and interface arts. They also need a new name.

The Rise of Urban Informatics

Just thirty years ago, "smart city" meant "fashionable dress."[3] Just ten years ago, "smart grid" had yet to appear in the mainstream news media.[4] And, less than a decade ago, the field of urban informatics first emerged. In 2006, the U.S. technology research journal *IEEE Pervasive Computing* organized a conference theme on *urban computing*, a term introduced by Eric Paulos, then at Intel Research.[5] The following year, a workshop on "urban informatics" was held in Australia, and a research handbook by Marcus Foth published.[6] The coinage "urban informatics" is often credited to the pioneering virtual communitarian Howard Rheingold, who foresaw the significance of street-level experience to digital culture.[7] Rheingold was responding to the New York City Wireless Initiative and to the writings of William J. Mitchell, whose urban technology trilogy told of "teleserviced neighborhoods" and "computers for living in."[8]

For architects, "smart city" means a departure from the algorithmically fabricated forms that have preoccupied most digital designers; for engineers, it represents a departure from all-powerful handheld gadgets. In a 2006 interview with *Metropolis* magazine, Mitchell explained how the smart city is not all mobile; it also runs on new combinations with embedded intelligence: "A particularly powerful design strategy under these conditions is to look for the ways that embedded intelligence loosens traditional relationships and constraints, and seize these as opportunities for fundamentally reimagining a product or system's organization, shape, and scale."[9] Thus the oft-cited MIT project for a new CityCar applied the battery capacity from racks of parked cars to citywide power storage balancing. Even everyday Zipcars demonstrate the network principle of product-service systems, nontragic commons, and productive combinations of mobile and embedded technologies.

"The real-time city is now real!" declared MIT's SENSEable City Laboratory in 2011. "The way we describe and understand cities is being radically transformed—alongside the tools we use to design them and impact on their physical structure."[10] Among the lab's many well-known projects, the Copenhagen Wheel (2009) combined energy harvesting, route selection, and ambient environmental data for bicyclists. After an interview with lab director Carlo Ratti, blogger Dan Hill described a "new soft city," where "you can see real-time information along one slice, one axis, and this enables us to anticipate a future city where perhaps the majority of the urban activity will generate impossible swathes of real-time data."[11] In his 2010 book, *Smart Things*, Mike Kuniavsky connected this phenomenon of "information shadows" with a more fundamental notion of "information as a material." Both are evident in street-level resources such as Zipcar, or its bicycle counterpart, Velib. "Information processing no longer needs to be the purpose of an object, but is one of many qualities that enables it to be useful and desirable in ways that are more directly related to people's wants and needs. In other words, information processing no longer defines the identity of an object, but is one of many materials from objects can be made."[12] To the visionaries of urban informatics mentioned here, these new materials and shadows become as intrinsic a part of embodied urban experience as tags, city lights, and media facades.

Over the last decade, hundreds of aspiring labs have produced thousands of street-level applications for arts festivals like ZeroOne and Ars Electronica. Research conferences such as those sponsored by *IEEE Pervasive Computing* increasingly accept smart city design project presentations. Burgeoning business conferences such as Where 2.0 test the entrepreneurial prospects of street-level location-based media. Interlink research policy initiatives from the European Union focus on "ambient computing and communications environments."[13] Big technology corporations have entered the field as well. IBM, for example, now promotes "A Smarter Planet." In a white paper entitled "Smarter Cities for Smarter Growth," IBM asserted the importance of better information services to overall urban prosperity. The experience of using urban infrastructures has become an ever more crucial component of livability, as measured by, say, the Human Development Index. Seen from the top, where IBM provides consultation to policy makers and infrastructure builders, the challenge is to integrate. The city is a "system of systems," which integrates core services in transportation, health care, public safety, and public education. But even from the top, this challenge increasingly emphasizes bottom-up social phenomena. The way to integrate, the white paper asserts, is to leverage the vast amount of existing data that accumulates in the course of everyday behaviors, and to make it "widely accessible to citizens."[14]

"Smart Grid will be bigger than the Internet,"[15] Cisco's CEO John Chambers proclaimed in 2010 as his company joined the race to build new energy infrastructure. Pervasive computing pioneers have often pointed out that, like electrical power in the twentieth century, digital processing in the twenty-first has disappeared into everyday life. Arguably the core technology of modernity, electricity introduced such concepts as appliances, pay as you go, and the grid itself into popular consciousness.

Alas, the electrical grid suffers from excessively top-down control, with huge power plants and distribution networks administered as public utilities; and it has been astonishingly wasteful, not just in how it transmits power but also in how its end users apply that power. By many estimates, a third to a half of the electricity used in buildings in the United States is wasted, and, by most estimates, buildings surpass vehicles as producing the largest fraction of the nation's avoidable carbon emissions. Thus electricity now seems ripe for, as Internet strategists would put it, "distributed social production." Today's investors bet on smart grids; consumers become cogenerators; devices time their operations to help balance demand loads; lights turn themselves off when you leave the room; and organizations actively monitor and reconfigure their consumption patterns. And, as with electricity, so with many other aspects of everyday life.

A more bottom-up approach to smart cities presents a new kind of design challenge. Just as electrification in the early twentieth century gave rise to a new discipline of industrial design, so smart, distributed, interoperable, data-intensive, citizen-accessible urban infrastructure in the early twenty-first is giving rise to a new discipline of pervasive interaction design. "Street computing" provides another possible name for this shift. As explained by Marcus Foth, who has organized events and publications under this name, street computing at its core facilitates better bottom-up awareness of the city, making more systems queriable and programmable.[16] As with electrification, this enables unforeseen appropriations and engenders new kinds of participation.[17] In the words of Eric Paulos: "We need to expand our perceptions of our mobile phones as simply a communication tool and celebrate them in their new role as personal measurement instruments capable of sensing our natural environment and empowering collective action through everyday grassroots citizen science across blocks, neighborhoods, cities, and nations."[18]

This participatory information stewardship transforms perceptions, both individual and social, of the city itself. Then, as urban usability constructs agreements to participate, to monitor, and to seek stewardship, it begins to take on aspects of a situated information commons.

A New Mental Map

With urban computing, "psychogeography" has entered a different era. Relations between embodied cognition, spatial mental maps, and explicit wayshowing systems now slip apart and recombine. From the perspective of architecture and urbanism, street-level media increase the importance of having worthwhile places to go. From the perspective of habitual attention, "worthwhile" means something more than momentarily amusing. In the rise of urban informatics, active participation supplants passive amusement (figure 9.1).

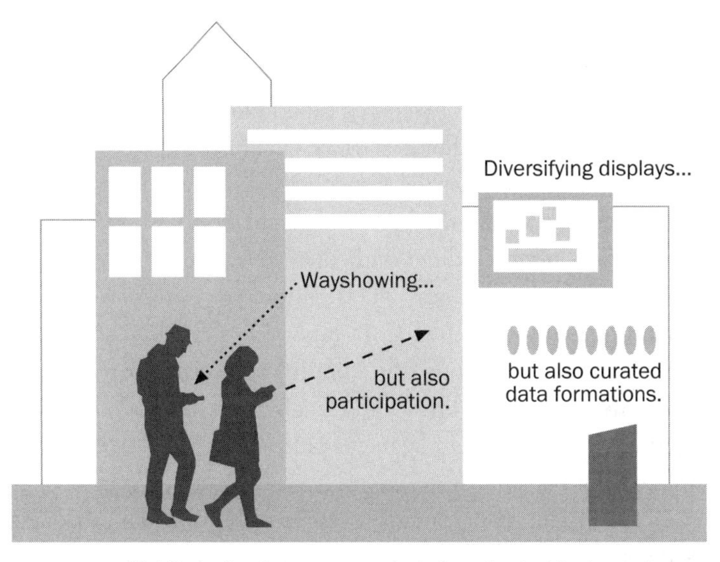

9.1 The basic idea of urban informatics.

"Psychogeography" was coined by the mid-twentieth-century situationists, whom today's proponents of situated technology still read. Reacting to the politics of the broadcast monoculture, under which they saw the terms of viewing increasingly being furnished, the situationists proposed that the best way to step out of that monoculture was to engage physical space in unanticipated terms, The best way to do so was to walk a playful drift (*dérive*) among less-noticed things, to bring some of those things into telling juxtapositions (*détournement*) that would break the spell of the politically engineered distractions (*spectacles*). To a situationist, play does not mean games with fixed rules so much as crossing in and out of states of expectation. This works better through serendipitous choice of which circumstances to ignore and which to exaggerate than by retreating into declared sets of personal preferences.[19] To the best-known situationist, Guy Debord, who is most often credited with these terms, psychogeography cultivated a self-awareness of attention to surroundings. By means of playful departures from expected behavior, whether for personal or social reasons, the mindful citizen could repurpose situations, and so reveal how those engineered distraction.[20] This mindfulness had ambient character; Debord referred to it as "the ambiance of play."[21]

Today's technologies differ substantially from the broadcast media of the situationists' era, of course. The monoculture that the situationists protested has dissipated. As noted in the chapter on screens, media have proliferated to the point where planned spectacles go unnoticed. The capacity to create spatial mental models hasn't really changed, however. The body imposes a schema on space, and the arrangement of bodies in space expresses those schemas in society. Tacit knowledge of these configurations informs spatial mental models, whether of communities of practice, contested ground, or anonymous drift along avenues.

Also, as noted in the chapter on embodiment, elements of mental models become internalized and externalized by activity. According to first principles of anthropology, the experience of urban activity emphasizes interpersonal distance, spatial distributions of hierarchical orders, and sites of collective commemoration.[22] Landmarks, districts, edges tacit and explicit, and nodes among one or more infrastructures provide the building blocks of spatial mental models. These models often take the form of ever-adapting collages of such elements, and seldom occur just as tags on a two-dimensional projection like a Google map, or any single uniform Cartesian view. The most famous visualizations of these models might be the "cognitive maps" created by architects and planners in the 1960s and 1970s, the most original of which was the oft-cited "Image of the City" by urban planner Kevin Lynch.

Urban exploration applications in mobile and embedded computing should thus be of considerable interest to cognition researchers. Where does the augmented city amplify the advantages of embodiment, and where does it cancel them out? When mediation such as GPS increasingly assists externalization, what happens to internalization? How do social navigation and more overt declaration of interests and preferences reshape street-level serendipity? If, after a decade of street-level urban informatics, everyone were to put their technology away, would their city skills be higher or lower than they had been before the technology? Or does the infusion of space with so many media simply erase all spatial mental models? Does the covering of high-resolution intrinsic information with lower-resolution processed information reduce affordances or affinities for embodied cognition overall, and thus reduce the image of the city as well?

You might expect that personal choices about maintaining a sensibility to surroundings figure in this. The influence of technology on urban experience might depend on your attitude toward environment, information as a material, or perceptions of overload. All of which makes universalist media and their sociologies suspect.

You might also read the paradigm shift from virtual world-building to urban informatics as an optimistic indicator of continuing spatial affinities. The exercise of embodied cognition can be restorative. It can feel more natural than purely abstract symbol-processing skills. Urban informatics can tap latent spatial abilities. To Carlo Ratti, this makes it "more Spacebook than Facebook."[23]

In the principles of embodied cognition, participation itself is situated. Street computing doesn't simply add a layer of portals to someplace else, but instead adds to cognition of the present place. It doesn't command attention on one channel at a time, but instead interleaves media objects among themselves and with unmediated objects, and in effect becomes ambient. Sites, props, social contexts, and interpersonal protocols of conduct produce a sense of engagement, which surpasses solitary use of a handheld device on a universal network at providing a sense of belonging, learning, or craft. According to philosophers from many different ages, those habits of skilled, purposeful engagement make better citizens.[24]

The casual, provisional arrangements of everyday life in the megacity remain elusive, however. Although the major builders of mobile and embedded technology have doubtless undertaken private studies of these arrangements, published studies such as the biennial working papers of Sri Lanka–based LIRNEasia on mobile technology practices at the bottom of the pyramid are few and far between.[25] A comprehensive street-level ethnography of media practices in the new megacities has yet to emerge. Because it would be difficult to find overarching unity in the currently sparse literature, for now, simply consider a few contrasting cases, particularly from the perspective of attention.

Wayshowing

When you combine a smartphone used as cursor with a positioning system such as GPS to look up nearby features, you get a "reality browser." To browse is to discover possibilities along the way; to browse reality is to combine the use of labels and links with presence in the physical spaces they describe. That can't happen in virtual spaces because a sense of presence depends on embodiment in haptic orientation and the inner ear. Now street-level media are available to help in the exercise of those. For someone who grew up being driven everywhere, street-level media may provide a necessary externalization, to be followed by internalization, of some basic city skills. This advances the centuries-old agenda of inscribing the city for incidental visitors.

Socially acceptable augmentations do exist. First off, most digital navigation is not to commercial offerings, but to friends. Social navigation now adds checking in to its moves. To declare your location on a social reality browser such as Foursquare lets unplanned encounters occur. To share tags and applications generates social life around particular activities and *dérives*, whether the active gaming of Parkour, the field identifications of plants or birds by naturalists, or the eccentric quests of collectors. Because a better wayshowing app makes systems of tags and labels available only to those who are interested, it helps urban explorers with filtering. The more that tags work as digital augmentations, the less they clutter physical spaces. On the other hand, such filtering serves to fragment the social sphere and creates new forms of digital divides.

Most people regard unfiltered, passive augmentations as little more than surveillance, which helps explain the generally negative view of pervasive computing. However, concern about an Orwellian Big Brother may overlook a more real concern about just how many thousands of little brothers are skimming personal data. Consumer analytics have moved beyond your desktop click stream to your physical movements in the built environment. Retail planning was already a science of positioning; and now advertising, the discipline most adept at media placement, may use proximity and spatial movement pattern recognition to deliver messages into contexts where they are more likely to be noticed. Tracking may also employ sensors, even face-interpreting software, embedded into aisles and shelves. Thus the Quividi audience measurement service uses visual analytics to document how long you look at a particular display.[26] Target audio beam technologies allow a spoken message to be delivered to a precise location when triggered by a motion sensor.[27] Abuses of attention rights may have only just begun. In other words, the prevailing early trends of urban informatics as wayshowing do not bode well for a tangible information commons.

New Epigraphy

Researchers and critics alike advance an urban informatics based on participation. As explored in the chapter on tagging, a new middle ground emerges between official inscriptions and transgressive graffiti, which could be called the "new epigraphy." New forms of annotation invite membership organizations, curation, and study.

Previous forms of signage have increased the usability of the city for the casual or unfamiliar visitor. But, for the resident, they are presumably unnecessary, and possibly an annoyance. The resident takes pride in awareness of changes to neighborhood amenities and the everyday routes they establish. The resident makes more use of intrinsic information and takes many more objects and events as signs. These are often of neighbors, of the encroachment of unwanted developers (who tend to trample on unquantified forms of local value), or of the need for civic services. Thus the highly successful maintenance wish list site SeeClickFix, which uses "citizen" prominently in its mission statement, rallies residents: that one resident expresses concern about an amenity lets another care, too. Other hyperlocal aggregators work across a variety of interests; outside.in, a pioneering hyperlocal news service, aggregates bloggers by location, and establishes a mood of curating local lore.[28]

Sound mapping works as urban storytelling, too. Tactical Sound Garden (TSG), another oft-cited project, demonstrated this process for the favored hotspot of Bryant Park, the birthplace of wireless Internet civics in New York City. Using three-dimensional positional technology, participants install a zone of audio overlays for browsing by anyone with headphones and a Wi-Fi device. Many such sound gardens develop on particular themes, such as local history, tagger culture, signspotting, or remembrance. TSG is currently an open-source toolkit for planning and "pruning" (modifying playback parameters) of sound gardens anywhere with good Wi-Fi coverage. A similar process works for images, incidentally. One famous Layar app lets you see images of the Berlin Wall in the context where it once stood, as shown in figure 9.2.

Much as networking has long allowed amateurs to become aggregators and producers of music and images, so now it allows them to gather environmental data. Thus, Living Light (figure 9.3) let participants text data to and from a park pavilion display

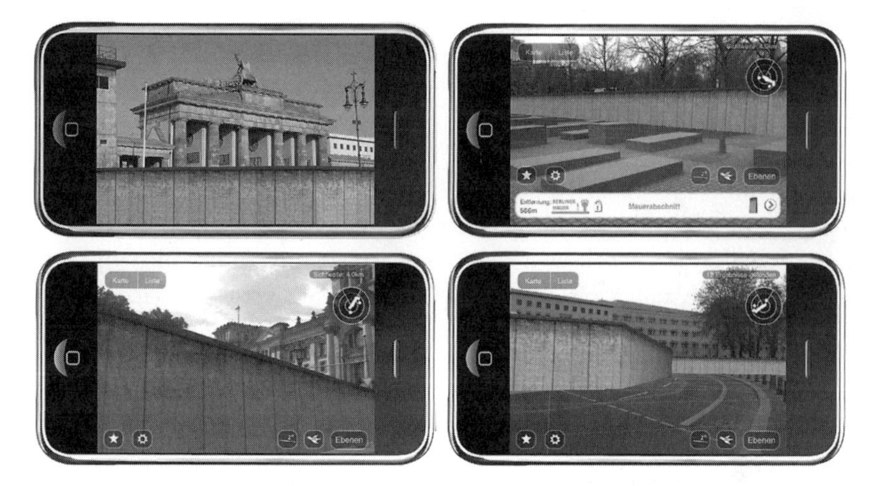

9.2 Reality browser: "Berlin Wall 3D" for Layar, Hoppala Agency, 2010. Photo: Marc René Gardeya, Hoppala.

of accumulated energy usage data, which compared the present year to the previous one across the Seoul region. Projects like this raise a very good question: how do cultural curators of participatory urban annotation systems see their work in relation to traditional or physical aspects of a commons?

In the recent compilation *From Social Butterfly to Engaged Citizen*, which includes cases on food, traffic, gardens, radio, crowds, and membership organizations, several leading scholars have offered positions on the ethics of urban social computing.[29] Many of the participatory qualities of Web 2.0 become more significant when coupled with the activities of daily life. For, just as the attention costs of passive media and autonomous annoyances are greater when you can't click away from them, so the

9.3 Open-source ecofeedback: Living Light, Seoul, The Living, 2008. Photo: Soo-in Yang.

benefits of active media and social networks feel greater when you apply them to shared physical environments.

Active participation in situated technology has most often taken the form of do-it-yourself (DIY) environmental monitoring. Participants sample, upload, map, and share data on pollutants such as carbon monoxide, surveillance cameras, invasive species, and noise. The Copenhagen Wheel project mapped levels of noise or air pollution by assembling geotagged data sampled by bicyclists as they moved around town. In an earlier instance of distributed sensing, Pigeon Blog (2006), took air samples from gas sensors and GPS readers attached, like paper messages of yore, to the legs of carrier pigeons.[30] Many such DIY monitoring projects now exist. "Turn your mobile phone into an environmental sensor and participate to the monitoring of noise pollution," invites NoiseTube, a Paris-based initiative sponsored by Sony.[31]

The use of personal communication devices to monitor, mix, and redistribute environmental data has a better name than "urban informatics," namely "citizen science." Eric Paulos, Ben Hooker, and R. J. Honicky introduced this term as an expression of empowerment.[32] Phones become data instruments; streets become platforms; aggregations become open-source communities, such as the data infrastructure platform Pachubé (now Cosm). Reports and displays become public embellishments, often in ambient format, such as the data murals of water and energy usage in the Arup Bangaroo project in New South Wales, Australia, that Dan Hill helped produce. Citizen science, then, is a use of technology for tuning in rather than out. Urban computing becomes alertness, perhaps even resilience, and not mere entertainment. Paulos and colleagues assert what this is not: "Urban computing is not a disconnected personal phone application, a domestic networked appliance, a mobile route planning application, an office-scheduling tool, or a social networking service."[33] Thus it is also not just for casual outsiders, such as tourists and advertisers. It is not personalized, nor is it as usable for beginners as it is for masters. With citizen science, new genres of urban data curation such as urban computing become a significant cultural domain.

U-City

Alas, too much occurs from the top down. Much of the rapidly urbanizing world is not so historic, civic, and complexly layered as a New York– or London-based flaneur or citizen scientist might wish. In most any metropolis, the walkable core that attracts global business and tourism is surrounded by a far larger ring of less centralized conditions. For example, many of the huge estates going up outside Beijing are designed to manage as many of their own infrastructural needs as possible. Many of the smart cities that get so much airplay from their corporate stakeholders constitute a case as different from core-city wayshowing and neighborhood citizen science as these are from one another. Layering means less in places where nothing preexists and where the technological emphasis is on ubiquity instead of augmentation. Hardly the sites of postmodern cultural juxtaposition, these smart cities are exercises in late modernist cybernetics: digital nervous systems of command and control. Resource economics dictates this more than any drive toward political aggrandizement. A truly smart city would reduce some of the biggest logistical sources of carbon emissions, material waste, time delays, and emergency unpreparedness.

South Korea regularly lays claim to being the most networked nation on earth, as measured in rates of usage, availability, and capacity of its communications infrastructure. Seoul has demonstrated many everyday situated technology practices already: it is a leader in the use of QRC tagging; its Galleria West facade (2004) was among the earliest instances of programmable media facades. In remarkable contrast to the Parisian kiosks of the Victorian era (figure 9.4a), Seoul's Gangnam District "media poles" (2009; figure 9.4b) provide new instances of the sidewalk arts and events kiosk, now in LED technology, at obelisk scale, 12 meters (40 feet) tall. So it was reasonable for South Korea to be the first country to lay claim to the expression "u-City" (with "u" for "ubiquitous"). According to Jong-Sung Hwang of the

9.4 Street kiosks then and now.

(a) Paris, ca. 1880, in painting by Jean Beroud (Walter Art Museum/Creative Commons).

(b) Gangnam District media poles, Seoul, seoulspace, 2009.
Photo: courtesy of seoulspace.

National Information Society, no less than twenty-two u-City projects were under way at the height of the boom in 2007.[34] Korea's leadership in infrastructure, its need to balance the boom of Seoul with development elsewhere in the country, and its centralized practices of construction consortiums combined to push ubiquity as an aspect of cultural identity. In "Living on a Platform," a survey of smart cities in 2010, the *Economist* led with New Songdo City, the most-cited instance of full-scale cybernetic city building to date. In a Shanghai World Expo exhibit that year, the networking giant Cisco demonstrated "all the digital plumbing" underlying Songdo. For example, in a mockup command center, "visitors were given a demonstration of how city managers would react to an accident on a city-centre bridge: cameras zoom in, an ambulance is dispatched, traffic is rerouted to other bridges—all automatically, within seconds."[35]

Today, post–economic crash, Songdo sits less than halfway completed, a self-evident critique of top-down urbanism from the standpoint of bottom-up arts and sciences. The everyday online media are filled with outpourings on Songdo and its ilk. Masdar, the United Arab Emirates project for a top-down smart green city, is mocked for being built on oil revenue and surrounded with the shantytowns of those who built it but can't afford to live there.

To anyone without a stake, such digital utopias may seem like technology for technology's sake. Or worse, a smart city can be a perfect dystopian union of technology, capital, and distracted urban subjectivity. So, by contrast, consider the introduction of media bottom up, on the other side of the economic and cultural divide.

Telecenters

Much experience of infrastructural access occurs in circumstances directly opposite those of any u-City: bottom up, undercapitalized, mostly undocumented, relatively low tech. Although access in these circumstances receives less systematic study, it needs to be recognized for its potential.

In particular, the mobile phone has brought network experience to far more city dwellers than any other technology. The economic advantages of connectivity, findability, and location may be all the greater to those with no other information infrastructures at hand and with no prospects for top-down investment or appropriation. Anthropologist and photoblogger Jan Chipchase has explored how this new bottom-up layer, often the first information infrastructure in a locale, relates to other resource networks and how it can show privileged digerati ways to develop without imposing technology for the less wealthy.[36] The models of use are not those of consumption, hurriedness, ubiquitous service and support networks, or the presumed constant need for entertainment. And, in contrast to the dislocated experience of developed global cities by digital nomads such as Chipchase himself, these bottom-up patterns can only be situated in material circumstances and the intrinsic information of things.

Consider the case of "tap attendants," who wait by an intermittently active standpipe for the water to run, and then manage the queue of customers with buckets to be filled, charging each customer a small fee. Nabeel Hamdi, a leading voice on participatory urban development, has remarkable stories about these everyday infrastructure workers, who are often children. In a corrupt situation, an attendant might pay the city a certain amount from the fees collected in order to receive water at this standpipe only on his or her watch, and then surcharge customers for a more predictable outcome. Another such tap attendant service on the rise is the recharging of mobile phones, for a fee payable in minutes of phone service.

One way to avoid such gatekeeping is to put the infrastructure access in the open, usually through what are called "telecenters," with governance bottom up among existing neighborhood or village councils, as in a commons. India began to install such telecenters around 2000, under an initiative named "Gyandoot," in which a pilot project set up about forty of them, some as storefronts and some as roadside kiosks, each designed to serve a dozen or more nearby villages. The project received democracy-and-technology awards internationally. Soon the market took the kiosk telecenter format to a larger constituency. By 2007, there were some 6,000 e-Choupal telecenters in India. Whereas the state centers served mainly to get government information out to the villages, the private centers were more often operated business to business, sometimes as commons, especially for the benefit of farmers.

Operations and practices of the telecenters reflected the status and practices of appointed villagers. Terms of engagement differed from place to place and often improvised metaphors and tokens of use that were quite outside technologists' expectations. For, as Paul Dourish and Genevieve Bell have observed, this fresh "experience of infrastructure" reveals patterns of culture: "We refer not simply to physical infrastructures but more broadly to infrastructures as fundamental elements of the ways in which we encounter spaces—infrastructures of naming, infrastructures of mobility, infrastructures of separation, infrastructures of interaction, and so on."[37] The telecenters revealed the complementary nature of technical and social patterning.

When it comes to the economics of attention, megacity resource networks behave quite differently from more familiar patterns of media consumption. For one thing, there must be intrinsic information in an urban resource commons, as well as social sensemaking and physical mise-en-scène. Neighborhoods that are undercapitalized fiscally may use new networking technologies to apply other, nonfiscal kinds of capital, such as cultural customs of access and use. Thus the kiosk telecenter format has been put to use by larger organizations such as the housing rights coalition Slum and Shack Dwellers International (SSDI), which now operates in thirty-three countries. In contrast to the "bottom of the pyramid" metaphor used by market analysts, which presumes that higher outside forces will be the main instigators and beneficiaries of resource schemes, this networked commons metaphor presumes that millions of local organizations will uphold locally intrinsic value better, and thus provide advantages that larger, more remote markets and states simply can't.

Well-meaning outsiders need to research such topics in greater depth, but they also need to proceed with caution. The cultural workings of attention may reveal which technology appears advantageous, and which is merely a contrivance. Whereas, in the most highly developed cities, there is a danger of romanticizing the technology, in the less developed ones, there is a danger of romanticizing the anthropology.

Urban Resource Partnerships

Can urban computing lay the cultural groundwork for other tangible information commons? What would it take to spread best practices into more cities, across more social divides, and into more resource pools? Even the small set of cases here suggests a larger prospect. Urban resource partnerships take on aspects of commons. As the economist Elinor Ostrom explained: "The key to a more effective [commons] model is to encourage self-organized contracts between local participants in context."[38] As the street-level media pioneers Julian Bleecker and Nicholas Nova have explained, the patterns of use that gather around shared streams of public environmental data make them into tangible social objects that are more accessible to casual social attention.[39]

Cases already exist in water quality, biodiversity, energy leaks, and the right to see the dark night sky.[40] Economists of networked social production have shown how nonmarket, nongovernmental organizations can help realize the value, and not just the fiscal value, of hyperlocal resources, and not just material resources, but also the kinds measured by the Human Development Index.[41] The dynamics of housing, water, power, transit, currency, opportunity, expertise, public health, and environmental health—these have become the agenda in urban computing.

How the ambient truly becomes a commons may take a lifetime to discover. Some already apparent aspects of the way forward, including changing notions of commons itself, deserve more inquiry in the chapter ahead. But before turning to that, another, perhaps even more fundamental aspect of urban computing as psychogeography deserves emphasis here. After all, media do not simply annotate a preexisting city but also help create new understandings, uses, and tacit geographies of the city. So this is really a question of attention to surroundings, and that is a fundamental theme in urbanism.

Distraction Reconsidered

This inquiry into attention gains perspective from an environmental history of information. In an age of embodied information, seen here from the perspective of participatory urban computing, age-old expectations about distracted urban life my no longer seem quite so accurate. There has been a change in the nature of distraction.

Although it may always have existed, and by now the advertising industry has made it seem nearly universal, an attitude of distracted irreverence once was less usual, and the topic of a new sociology. Scholars of a mindful, resistant urbanism still recite Georg Simmel's 1903 portrait of distraction, "The Metropolis and Mental Life," in which "there is perhaps no psychic phenomenon which is so unconditionally reserved to the city as the blasé outlook."[42] Presciently, but not so uniquely, Simmel saw money steadily replacing all other forms of social exchange (a process that continues today in what social media tycoons now call "monetization"). Like other early sociologists, he saw a steady decline in everyday opportunities for spontaneous personal engagement, as city dwellers dealt more with strangers, identified less with groups, spent much more time alone, and worked as cogs in some giant machine.

For as postmodern critics so often protested, visual culture itself industrialized; and in the process, so did attention. The interplay of distraction and attention only took modern form in the last third of the nineteenth century. That is when William James began to explore it, for one. Industrialization had made attention into something to pay, not only when attending factory machines, but also with respect to visual culture. As art historian Jonathan Crary observed, "modern distraction was not a disruption of stable or 'natural' kinds of sustained, value-laden perception that had existed for centuries but was [instead] an effect, and in many cases a constituent element, of the many attempts to produce attentiveness in human subjects."[43] Through careful reading of both early texts of the then-formative discipline of psychology and selected paintings from the period, Crary was able to identify attention as a new idea. "Not until the 1870s does one find attention consistently being attributed a central and formative role ..."[44]

In what became his more lasting, unique contribution, Simmel reacted against this new sense of attention. Whereas "anomie," introduced by his more influential contemporary, Emile Durkheim, conveyed a general sense of disconnected outlook, "blasé" and its English equivalents "blunted" and "dulled" expressed it in more personal, perceptual terms. In a fittingly industrial metaphor, "blasé" means worn down through excess, not only from the labor or pollution that many sociologists protested, but also from unprecedented diversity of demands on attention, or as Simmel put it, "incapacity to react to new stimulations with the required amount of energy."[45]

This incapacity arises from the need to shift attention quickly and often. In what may be the most famous passage from "The Metropolis and Mental Life," the fatigue that dulls and blunts comes from "the intensification of nervous stimulation, resulting from the rapid telescoping of changing images, pronounced differences in what is grasped at a single glance, and the unexpectedness of violent stimuli."[46] Or, in another translation, it results "from the rapid crowding of changing images, the sharp discontinuity in the grasp of a single glance, and the unexpectedness of onrushing impressions. These are the psychological conditions which the metropolis creates." Long before handheld communications, outdoor video, or electronic ink, the flood of stimuli was enough to make distinctions among its elements vanish, giving rise to city dwellers' characteristic "blasé attitude," whose "essence" Simmel described as "an indifference to the distinctions between things."[47]

Although Simmel's larger work on political economy has been largely forgotten, his particular focus on dulled subjectivity eventually resonated with the late twentieth-century critics, who revived him.[48] As consumerism reached unprecedented levels in the 1980s, Simmel seemed far ahead of his time on the experience of fragmented, decontextualized, desire-inducing media. Postmodernists found Simmel's essayistic, anticomprehensive style appealing.[49] For, as they would have put it, the blasé privileged the reader. They agreed how the response of city dwellers to the readymade life, its furnished worldviews, and its endless overstimulation, was to become highly arbitrary and distinct in one's tastes.[50] The unprecedented material benefits (electricity, sanitation, transit, communications) that modern cities provided their citizens made that possible.[51] Although distraction and overload could occur in any culture, modernity offered more means to become comfortably numb. Or, in Simmel's words: "as a protection of the inner life against the domination of the metropolis, the reaction of the metropolitan person to those events is moved to a sphere of mental activity which is least sensitive and which is furthest removed from the depths of the personality."[52]

Today, the onrushing impressions have become more numerous, more subtle, and more widely distributed than in Simmel's time. This is the usual qualification that twenty-first-century critics make to the argument that people have always experienced overload. Yes of course they have, but not so often, not in so many different aspects of everyday life as now, and not by such easy means. The harsh industrial distractions of city life have waned; there are fewer things belching steam, soot, and noise at such intensity. Today, much more in the flood of stimuli takes the form of intentionally produced, subtly appealing or entertaining, widely distributed media productions. You may experience ever more of these productions involuntarily, in part because they so pervade the activities of your lives that despite all diligence you cannot keep up with the filtering. But then, more significant to this inquiry, the flood of stimuli also occurs at street level, where it is even more difficult to escape.

In short, never has distraction had such capacity to become total. Enclosed in cars, often in headphones, seldom in places where encounters are left to chance, often opting out of face-to-face meetings, and ever pursuing and being pursued by designed experiences, postmodern posturban city dwellers don't become dulled into retreat from public life; they grow up that way. The challenge is to reconnect.

Meanwhile, the experience of information overconsumption has developed a much more participatory, social infrastructure. Simmel was witnessing the rise of one-to-many commercial media, albeit before electronic broadcast technologies brought them to the center of everyday life. The postmodernists who revived Simmel were witnessing the absurd extremes at the end of one-to-many media dominance—the 1980s were the last decade of television monoculture. And the urban computing pioneers who today translate an interest in Simmel forward to the age of personal street-level media are witnessing the rise of many-to-many, or what some call "read/write" urbanism. Where an ethics of street computing engenders citizen science and notions of commons, the microstructure of engagement stands in dramatic contrast to the disengagement of city dwellers dulled by mass media.

In sum, a different sense of overload seems inevitable at each different stage in the history of environment, information, and technology. To someone displaced from traditional rustic life, where that tradition seems recent and memorable enough for constant comparison, urbanism amplifies the sense of displacement, or anomie. To someone who grew up in postindustrial sprawl, with disembodied friendships, nonstop media feeds, and informational empty calories, urbanism represents a prospect for relative sanity, or at least a richer mix of perceptual options, and a better balance among information about, for, and as the world.

This voluntarily urban citizen prizes attention skills, defends attention rights, and takes time for attention restoration. And that seems quite different from sitting alone, grazing on favorite feeds, and hoping not to miss any messages. It also seems different from Simmel's shock at the newly electrified Berlin. Overstimulation may be more subtle, widespread, and appealing than before, but blasé has become less of an option. Those who go blank become only more vulnerable to thoughtless overconsumption, even attention theft. Instead, the best defense is to choose to take interest, and to help your sensibilities slowly evolve.

How newer megacities now urbanize will have more impact than what the existing metropolises do next. This process is much more difficult to study, to capture with art installations, or to read or write books about. Millions of people now network their local resources, organize governance where markets and states have missed doing so, uphold nonfiscal capital in nontraditional ways, use embodied media to form their images of the city, and so recast their workings of attention.

9.	MEGACITY RESOURCES
Main idea:	Urban computing inevitably transforms attention to context
Counterargument:	Don't impose technologyKey terms: Urban informatics, psychogeography, resource networks
What has changed:	Bottom-up economics of rapid urbanization
Catalyst:	More kinds of resource organizations
Related field:	Smart cities
Open debate:	Non-market networked production?

10 Environmental History

Information deserves its own environmentalism. The more that information technology permeates everyday life, the more inescapably it alters personal and cultural sensibilities. Of course, the physical patterns of everyday life can be just as telling as a culture's art or politics. Thus, one culture, whose citizens variously walk, ride bicycles, drive cars, and take streetcars to get from place to place, might assume they need little instruction to share the streets, whereas another, whose citizens almost always move around in cars, might need plenty of signage, and might sometimes use parking restrictions to avoid unanticipated social mixing. To understand such cultural differences, it can help to see their many usage patterns as a landscape. It can also help to see cultural landscapes in historical perspective. It can help to see such larger patterns as "cultural landscape." In a widely-read definition of landscape, the design critic Paul Shepheard once advised that "the big moves in [a cultural] landscape happen very rarely. You will be lucky to see one during your lifetime and even luckier to be involved in the making of it. Incremental changes happen all the time, however."[1]

Today's big moves are happening as humanity rethinks its role on earth. Countless incremental moves contribute to that epochal change, and in the process transform the relations of environment and technology. Although mechanistic modernists dismissed it as romanticism and technofuturists dismissed it as irrelevant to a disembodied cyberspace, today environmentalism has become the most practical attitude of all. The word *landscape* had its origins in naturalistic romanticism. One way to outgrow that attitude is to understand many more contexts as "environment"[2]—not just places untouched by technology, but also buildings, cities, and their diversifying layers of information.

Disposition, History, and Criticism

Few words have become so overexposed or misused as *environment*. So long as it meant someplace large and far away, outside of technology and beyond human agency, the "environment" was difficult to imagine, much less to engage. And so long as information technologies were considered in the abstract—having neither place nor substance—they could hardly shape an environmental imagination. That is now changing. Information technology builders are devoting more time and attention to environment; and environmentalists, for their part, are finding more to like about pervasive computing. As the conceptual gap between technologists and environmentalists closes and critical projects in smart green technology become realities, it is worth sorting out environmental terms as they are now used.

Environmental disposition describes priorities and perceptions and implies both receptivity to figurative objects and events and awareness of the fields of space they share. That awareness might be visual, an awareness of the special light on Penobscot Bay that has attracted artists to Maine for generations, for example. It might be territorial, an awareness of the lay of the land, long and acutely cultivated by the military, or of city markings as these express tacit geographies of neighborhoods and gangs. Or it might be mythological, an awareness of enchantment in landscape among members of ancient cultures. In an industrial society, however, any environmental disposition is likely to be countercultural, although those born into an age where environmental crisis is an ongoing fact of life might have difficulty imagining how recently this was so.[3]

The character of "disposition" is more substantial than it might first appear: not only do the means of production and notions of nature differ significantly from culture to culture, but so do the plant and animal species, the droughts and pathogens, and the shape of the land itself. In his 1997 book, *Guns, Germs, and Steel,* biological historian Jared Diamond showed how the Anglo-American mythology of "Manifest Destiny" in the overall westward course of empire was based in biological differences among continents, shaped by their physical geographies, rather than in race.[4]

Environmental history explores a two-way relation between cultural disposition and land transformation. To some historians, that modern California got its start in hydraulic methods of gold mining, which cultivated technical ingenuity and capitalization, helps explain why Silicon Valley is there and not somewhere else.[5] As with so much else in early environmental work, historians simply tried to document environmental impact, often beginning with political economy. Thus a slash-and-burn agrarian culture and one whose members build irrigation canals they must share and maintain are likely to have quite different worldviews. (That an expression like "slash-and-burn" has become a widely used metaphor in other disciplines attests to the growing reach of environmental thinking.) As first-generation environmental historian Donald Worster once declared: "In the old days, the discipline of history had an altogether easier task. Everyone knew that the only important subject was politics and the only important terrain was the nation-state."[6] To study production was a change. The origins of the disciplinary shift have generally been credited to the mid-twentieth-century French *Annales* school of economic historians. To make their discipline distinct, early environmental historians excluded buildings and artifacts from their scope.[7] Only more recently has the now-established discipline accepted the need to look at material circumstance in order to question cultural assumptions. The politics exposed are, more often than not, the mechanisms of everyday consumerism. Which patterns of life exist out of raw necessity, and which out of cultural assumption? Feedlots? Lawns? Green second homes? Nature stores?

Normally, the word *criticism* arises in arts and letters, and has implications of taste. Its role of interpreting expression distinguishes it from mere complaint. Criticism implies the aspiration to improve indirectly by interpretation.[8] This differs from compiling so many random opinions, as some websites do. Critics must know not only what they like or dislike, but why. This comes from knowledge of a genre, based on appreciation of a medium's affordances and constraints, by which critics can compare works and see bodies of work.

The term *environmental criticism* was first advanced by literary historian Lawrence Buell. The mission of environmental criticism, Buell argued, was to help move the environmentalist agenda beyond protecting nature, beyond thinking of nature as independent of human settlements and morally superior to anything made by humans.[9] Changing naturalism was also a concern of the late Welsh novelist and critic Raymond Williams, whose works achieved that change through careful shifts away from romanticizing the country, as if it could not exist in its present form without the city.[10]

The term *environmental artifice* might seem oxymoronic when viewed through the simplistic opposition of nature (good) versus artifice (bad). But artifice shapes environmental dispositions, for better or worse. This could be as simple as crop irrigation and as complex as biodiversity upkeep policies. Although there are those who believe that any human intervention detracts from a place, as if the Golden Gate would be better without that big red bridge, designers must assume that appropriate intervention does overall good.

Among forms of artifice, the city has the greatest inhabitability and persistence. Although urban histories often suggest that geography is destiny, they mainly focus on city form itself; that is their environmentalism. More specifically, urban history looks for transformations in the conception of space itself through the evidence of city form, as when Paris, in the mid-nineteenth century, cut broad straight boulevards through its old (and rebellious) medieval quarters. Urban history tells stories of infrastructure, whether sewers, railroads, or radio stations. It tells how the latest phases of technological development more often overlie their predecessors than eliminate them. It looks at repeating elements such as housing types and their relationships to resource flows, whether of trains, water, or communications. On a larger scale, it studies the relationship of cities to their hinterlands. In a genre-changing turn, William Cronon's 1991 classic, *Nature's Metropolis*, showed how the grain, meat, and timber markets of Chicago transformed a thousand miles of hinterland and how this led to specialties in artifice, such as mail-order catalogs and refrigerator railroad cars.[11]

The history of the city offers vivid small instances of situated technology, especially as fittings to larger infrastructures. Thus, among the many new practices that arose around electrified underground rail access to city centers, the scurry of commuters through Grand Central Station led to giant clocks, escalators, and fast food.[12] At the Horn & Hardart automat two blocks from Grand Central, commuters faced wall-sized arrays of coin-operated little glass window that opened onto plates of food.

Formerly separate disciplines of urban and environmental history first converged on the topic of pollution. Both traced how industrialization brought not only unprecedented squalor but also unprecedented wealth, part of which could be used to relieve the squalor. Industrial life has for centuries produced sludge and smoke, as well as advertising and noise. Doing anything about any of these side effects, however, is only a recent development. This has also been true of pollution in largely non-industrial societies such as India's, where the taboo against deliberate contact with profane waste resulted in the holy Ganges being ever more fouled with incompletely cremated corpses. Eventually, however, something had to done; the authorities introduced carnivorous snapping turtles to clean up the river.[13]

In his 1936 landmark book, *Technics and Civilization*, Lewis Mumford observed that "the first mark of paleotechnic industry was the pollution of the air." "Neotechnic" industry shifted

10.1 Scrubbing the soot from Pennsylvania Station, Pittsburgh, 1948 (Heinz History Center).

"from destruction to conservation" of the environment.[14] For example, in nineteenth-century Pittsburgh, smoke and fire were discussed aplenty, but to describe them as "pollution" was to think of them as correctable, a change in perspective that occurred only much later.[15] Famously, the Pittsburgh Survey of 1907 brought a progressive eye, and unprecedented levels of documentation, to the living conditions of immigrant labor in that notoriously smoky city. After World War II, Pittsburgh began to reduce and scrub off the soot (figure 10.1). Today,

Pittsburgh demonstrates how environment and livability can eventually become economic generators in themselves.

Today marks another new era in the history of the city. As noted in the previous chapter, the majority of humanity is now urban. While migrations of the rural poor to the megacities of the south have of course been the main drivers of this, note too a significant shift in the north, particularly in America. As the environmental, social, and public health costs of suburban automobile monoculture have become evident there, housing demand has shifted from exurban McMansions back to walkable urban neighborhoods.[16] Once the towering industrial city was the early environmentalists' epitome of what to avoid. Today, as made famous in David Owen's recognition of "Green Manhattan,"[17] walkable cities turn out to be a much greener choice than sprawling suburbs.

Walkability in dense neighborhoods happens to favor location-based social and curatorial media. As this inquiry has explored, urban computing has the potential to become an art form, a civic experience, and a new kind of information commons. From an historical perspective on the relationship of environment and technology, this invites a basic question. How can electronic artifice bring alive a sense of belonging to the world, and not just suggest conquest, distraction, or escape?

Environment in the History of Information

Histories of information, which became abundant during the web boom, emphasized long-term change in literacy, not in the environment. A comprehensive environmental history of information has yet to be written. As a way to begin, then, let us consider instances of environment in the history of information and of information in the history of environment, which seasoned historians are now exploring. Our present inquiry simply seeks to connect environmental histories with future prospects for augmented cities.

When the rise of the web challenged the dominance of print in the early 1990s, the history of information became a widely read genre. Much of this concentrated on the history of reading, which was then almost synonymous with print.[18] Twenty years later, after a great diversification of mobile and embedded technologies has infused information media into everyday life, there is little talk of dematerializing into cyberspace.

The history of information has far fewer documentary sources available to it than does the history of technology, say, or the history of everyday domestic life, or that of national politics. The kinds of circumstantial and ephemeral formats of reading that are of historical interest in this age of media diversification seldom received mention in the literary, artistic, or journalistic sources that historians typically mine. Indeed, the word *information* came into general use only in the nineteenth century.[19]

The rise of library science in the nineteenth century forced a conceptual switch from information as process to information as entity.[20] Alberto Manguel, in his well-respected *History of Reading*, observed how "the system Callimachus chose for [the great library at] Alexandria seems to have been based less on an orderly listing of the library's possessions than on a preconceived formulation of the world itself. . . . With Callimachus, the library became an organized reading space. . . . All the libraries I've known reflect that ancient library."[21] Manguel later wrote of his own private library in highly ambient terms, as an aggregate assembled by chance, a mental sanctuary, made to be sat in by night for its mood, a mood that could be felt even before reading any books.[22]

It is this reified form of information that was scarcer in preindustrial societies. We have always been informed of something, and empires have usually had libraries, but only in recent centuries have most of us had access to organized collections of informative texts—symbolically encoded semantic entities—newspapers, journals, and books. Each increase in this kind of information was without precedent; at any stage, it might have seemed like overload.[23]

"Contemporaries start to articulate the problem of the overabundance of books around the 1550s," Ann Blair has noted. "It does become a refrain."[24] One much-quoted line from the early centuries of print that Blair has studied is from the philosopher-mathematician Gottfried Leibniz, who complained about "that horrible mass of books which keeps on growing."[25]

The English country house, for example, has been studied by literary scholars for its social practices of reading.[26] It was here, in the seventeenth and eighteenth centuries, before significant revolutions in democracy and mass media, that the idea of abundant information began to take form. Here the opulence was not just palatial, but literary. Polite society expected gentlemen to be current on all the latest literature. Private libraries had come into existence, and this, too, distinguished the great halls from ordinary houses, many of which had only a prayer book or Bible, if any book at all.[27] Some of these libraries included new forms of reference materials that are of historical interest today. In a Jonathan Swift story from 1704, there is a much-quoted passage about the new practice of index skimming, which was done in order to proclaim familiarity with as much learning as possible:

"Thus men catch knowledge by throwing their wit on the posteriors of a book, as boys do sparrows by flinging salt upon the tail."[28] Such skimming was but a small part of the information experience of the times, however, which was expanding in a variety of ways. Postal traffic increased. Newspapers expanded in number, size, and frequency of publication. Ledgers and other business records (which in ancient times had been the first purpose of writing) became more substantially organized. Maps of faraway places held special interest, especially where these were the locales for business ventures.[29] In examining this diversification of print, and given how written communications were so often agents of change, the literary historian Katherine Elison has found information overload in the country house described as "fatal news."[30] Even as it began to be understood in itself, information was something overwhelming, even threatening.

To information historians such as Marshall McLuhan, this period of print diversification, and its role in the formation of modern capitalist nations, was just as revolutionary as the later electronic age. The many new forms of print later so basic to everyday city life first played a prominent role in the eighteenth century, where they transformed the experience of reading, and more specifically that of sharing. Histories of information often thus emphasize the social uses of print. More widespread reading transformed social and political life. Beyond the aristocratic salon and before the rise of public libraries (and the information sciences with them), many new venues and practices emerged, such as salons, shared private libraries, and clubs, like the famous Boston Athenaeum. To the skeptical, these were clear instances of overload, both for the increased reading duties and for how those cut into social manners (figure 10.2).

The Victorian city was awash in print like no city before it. With the telegraph to send news, but without the radio to broadcast it, this was the golden age of newspapers. Life became much faster, uglier, and more distracting. Many of the arts turned to realism; Emile Zola wrote vividly of all this in France. The print arts also increased their volume. In an oft-cited expression of glut, one follower of Zola named Octave Mirabeau wrote: "Books rise, overflow, spread, it's an inundation. From overcrowded bookstores breaks a torrent of yellow, blue, green, and red, cascading from displays that make you dizzy."[31]

Enough archival materials remain that historians may construct a view of nonbook text in the nineteenth-century city,[32] which invites contrast to that of hypertext in the twenty-first century. Although the word *media* had not yet come into common use, the word *public* had, thanks in no small part to the influence of newspapers. As more kinds of political talk could safely occur in the open, those who took part needed somewhere to gather casually. In contrast to the private country house, a new site of reading became much more readily available: the public coffeehouse. Here was a new kind of environment where the availability of information (although that word was not yet in use) was the reason for being. Instead of one person reading to many, or many all talking about one thing they had read, each patron sat reading something of his own (as women were generally excluded), and occasionally offering remarks from it to the others with whom he kept company. Although coffeehouses receive little mention even in the work of contemporary writers known for their detailed portraits of life, such as Laurence Sterne, Samuel Johnson, and their seventeenth-century predecessor, Samuel Pepys, modern scholarship has shown how print

10.2 Overload amid early mass print: *Everyone Reads Everything (Alles liest Alles)*, 1830, painting of a Berlin coffeehouse by Gustav Friedrich Taubert (Staatliche Museum Berlin/BPK).

communications became somehow experiential there, in ways aided by the coffee, of course.[33]

On the street, amid placard bearers and sandwich men, young boys pushed handbills about new coffeehouses into the hands of passers-by (figure 10.3). Others sold daily sheets of ship arrivals and market goods and prices. Earlier forms of these papers, named for the Venetian *gazzetta* coins used to buy them, have been identified as the first instances of print advertising, and as precursors to the industrial newspaper. Handbills and circulars passed from person to person about deals of the day became dominant forms of advertising.[34] Business cards were offered by those in particular trades, and calling cards presented for social purposes—elegant precursors to Facebook. The mail was delivered more than once a day in some cities, as were newspapers. The bustle of hand-delivered written communications is difficult to imagine in an age of instantaneous electronic surrogates. Only a few bicycle couriers remain, rare enough to be considered stylish.

Walls adjoining sidewalks were prime real estate for communications. A few surviving images of that era appear in today's civic campaigns against current flyposting practices that are but a shadow of what existed before glossy magazines, television, and Internet shopping. Posted bills seem distinctly characteristic to the street scenes of the early nineteenth century (figure 10.4), when color lithography had been developed, but not yet applied to mass-produced, domestically consumed magazines.

In his remarkable book on the written word in the public spaces of antebellum New York City, urban historian David Henkin has explored the significant contribution of casual, individually produced forms of printed papers to the urban

The Vertue of the *COFFEE* Drink..

First publiquely made and sold in England, by *Pasqua Rosee.*

THE Grain or Berry called *Coffee*, groweth upon little Trees, only in the *Deserts of Arabia.*

It is brought from thence, and drunk generally throughout all the Grand Seigniors Dominions.

It is a simple innocent thing, composed into a Drink, by being dryed in an Oven, and ground to Powder, and boiled up with Spring water, and about half a pint of it to be drunk, fasting an hour before, and not Eating an hour after, and to be taken as hot as possibly can be endured; the which will never fetch the skin off the mouth, or raise any Blisters, by re son of that Heat.

The Turks drink at meals and other times, is usually *Water*, and their Dyet consists much of *Fruit*, the *Crudities* whereof are very much corrected by this Drink.

The quality of this Drink is cold and Dry; and though it be a Dryer, yet it neither *heats*, nor *inflames* more then *hot Posset*.

It forecloseth the Orifice of the Stomack, and fortifies the heat within it's very good to help digestion; and therefore of great use to be bout 3 or 4 a Clock afternoon, as well as in the morning.

uen quickens the *Spirits*, and makes the Heart *Lightsome*.

is good against sore Eys, and the better if you hold your Head o'er it, and take in the Steem that way.

It suppresseth Fumes exceedingly, and therefore good against the *Head-ach*, and will very much stop any *Defluxion of Rheums*, that distil from the *Head* upon the *Stomack*, and so prevent and help *Consumptions*; and the *Cough of the Lungs*.

It is excellent to prevent and cure the *Dropsy*, *Gout*, and *Scurvy*.

It is known by experience to be better then any other Drying Drink for *People in years*, or *Children* that have any *running humors* upon them, as *the Kings Evil.* &c.

It is very good to prevent *Mis-carryings in Child-bearing Women*.

It is a most excellent Remedy against the *Spleen*, *Hypocondriack Winds*, or the like.

It will prevent *Drowsiness*, and make one fit for busines, if one have occasion to *Watch*; and therefore you are not to Drink of it *after Supper*, unless you intend to be *watchful*, for it will hinder sleep for 3 or 4 hours.

It is observed that in Turkey, where this is generally drunk, that they are not trobled with the Stone, Gout, Dropsie, or Scurvey, and that their Skins are exceeding cleer and white.

It is neither *Laxative* nor *Restringent*.

Made and Sold in St. *Michaels Alley* in *Cornhill*, by *Pasqua Rosee*, at the Signe of his own Head.

10.3 Surviving handbill from what is thought to be London's first coffeehouse, 1664 (British Library).

experience, amid the proliferation of official street signage and fixed commercial signs. Less like enduring inscriptions and more like speech acts, these more incidental writings were precursors of today's mobile texts, many serving as what sociologists now call "presentation of self." They publicly let others know of a social or business presence, and in doing so became part of an ever-changing aggregate layer of the city.[35] Conversely, this layer let strangers know that they had a right to be present. Only with the Industrial Revolution did most cities become large enough to be full of outsiders, immigrants, and transients. Until this happened, most locations, wayfinding, and local regulation could be adequately conveyed by custom or by word of mouth. This shift toward outsiders led to the modern sensibility of city dwellers and to the sense of distraction that so preoccupied later urbanists, such as Simmel. To Henkin, "we are both embraced and alienated; the city acknowledges and addresses us, but keeps us moving at an impersonal distance."[36]

Tourism takes acknowledgment of newcomers to an extreme. Anyone disoriented in an unfamiliar city becomes more reliant on environmental graphics. Then when visitors speak well of sites where the orientation process is pleasant, and where they find what they want, soon the traffic increases. This process economically rewards destinations that not only orient visitors but confirm their expectations. Thus, more than any other urban activity, tourism furnishes interpretations. This, too, has its origins in the early age of print, when guidebooks first appeared, as for Paris after the revolution.[37]

Like GPS today, this early wayshowing supplanted nontextual reading of intrinsic structure. As noted in the chapter on tagging, the architecture of the city aims to represent institutions (or other prominent parties) to their constituencies in ways that make sense without explanation. Architecture orients tacitly. Good city form is thought to be legible without interpretive signage. Tourists gravitate to such cultural landscapes because they feel better than most other, less orienting ones back home. Retail planners recognize that the highest returns per square foot come

10.4 Flyposting in its heyday: *A London Street Scene*, 1835, watercolor by John Orlando Parry (Wikimedia Commons).

from neighborhoods that are not only walkable but also memorable in form and identity. But, paradoxically, the economics of consumerism then layers and reconfigures these pleasingly memorable spaces according to their branded formulas, preempting and largely masking the intrinsic legibility that attracted visitors in the first place.

Historians of information thus need to expand the role of environment beyond the sites of reading, to reexamine the distributed application of information to the innate and built world. There are several reasons why they have been slow to do so. First, what we are calling "intrinsic information" is much more ubiquitous, innately reflecting the process of being in the world. Second, intrinsic information dominates traditional societies, where cosmology and mythology are among the main components of what modern societies might simply call "information." And, third, intrinsic information doesn't further an agenda of computation, whose rise has driven many quests for a prehistory of information.

Here is where the mainstream discipline of environmental history comes into play. As Carolyn Merchant, herself a prominent environmental historian, has explained: "Environmental history is both one of the oldest and newest fields within human history. All cultures have oral and written traditions that explain human origins and encounters with the natural world through stories about local landscapes and ways to perpetuate life from the land."[38] What the age of ambient information may learn from premodern histories of environmental information is how to read the world in itself.

Information in the History of Environment

Because environmental history examines human relationships with "nature," and because the arts and letters best reveal those relationships, it follows that most information in the history of environment takes the form of narrative descriptions, especially in myths and literature. Historically, especially in premodern cultures, a principal role for literary tradition is to enchant a people's place in a landscape.

Scholars, notably in art, linguistics, anthropology, political theory, and regional economic planning, have long studied these narratives and their graphic counterparts. Thus planning scholars have examined the impact of geographic information systems

(GIS) on cultural geography and land use patterns. More so than other, more ephemeral or poorly archived media such as printed broadsides or posted street signage, mapping demonstrates a long-term development of information media's environmental impact.

Throughout its long history, mapping has seemed inseparable from colonialism. To map a territory was to domesticate it, to name and measure a former terra incognita. Through the unique combination of democracy, relatively undeveloped territory, and emerging market forces, buying and selling lands became normal in early American history; for the first time, maps of these lands were owned and displayed by many members of society. David Hackett Fischer compared folkways of what were quite separate cultures in Massachusetts, Pennsylvania, and Virginia.[39] America was many separate peoples then, and each imposed its own particular cultural geography: the hub-and-spoke roads of New England, the small metes-and-bounds counties of Kentucky, and the later Jeffersonian grid from Ohio and Indiana westward.[40]

According to the main cultural narratives, not only was the land waiting to be developed, but this was its divine purpose and humanity's mission. Technology historian David Nye has examined assimilation of nature into human designs as the basis of several American foundation myths: "In the American beginning, after 1776, when the former colonies reimagined themselves as a self-created community, technologies were woven into national narratives. A few assumed particular prominence, among them the axe, the mill, the canal, the railroad, and the irrigation dam . . . with America conceived as a second creation built in harmony with God's first creation."[41] Few other cultures have made technology so intrinsic to their creation narratives.

The most visible case of second creation as an environmental consequence of an information practice can be seen best from an airplane: it is the Jeffersonian grid of the Midwest and Plains. Like the street names and signs of postrevolutionary Paris, America's 640-acre square of homestead land belongs to an era of standardizing measurement in the name of democracy, and to an agenda of bringing ownership to many more people. It illustrates how a change in the simple information practice of measurement can have huge consequences.[42] The immediate reason to survey land is to establish title, often for sale. The newly surveyed territories of the nation's Northwest became the site of speculative boom and bust just as dramatic as more recent Arizona subdivisions.

For a related example in the history of technology, perhaps the best retold tale is of railroads and their associated information media crossing these gridded prairies. The High Plains themselves have legendary status among environmental historians as one of the few places in the developed world that are depopulating today. Historians now marvel at railroad-based inducements to settle, such as the infamous earth-science category error that "rain follows the plow," or the distinct technological convergence of railroad deliveries with steel plows, barbed-wire fencing, and harvesting machinery. The railroads depended on their own supporting set of information technologies. Back on the streets of New York, handbills induced emigrants westward toward the bounties of Nebraska life. Along their routes, railroads had switch and signal systems of their own, necessary for better safety and operations, and precursors to later computational disciplines of queuing, from which the discipline of cybernetics emerged. Railroad corridors simultaneously served as telegraph trunk lines.

Cosmologically, these combined technologies "annihilated space," in the parlance of the day, and made necessary the systemization of timekeeping into discrete time zones, a loss of locality that was protested vigorously. Economically, although the railroads were by far the largest capital entities yet, countless smaller new business practices grew up around them, some built on new information practices, such as cash-on-delivery mail-order catalogs. In an architecturally distinct example of information technology in environmental history, the mail-ordered, flat-packed, designed-for-assembly, balloon-framed house gave a consistent look to a wide swath of new prairie towns.

For a twentieth-century case, consider radio as the first truly ambient information technology. Radio was the first "wireless" medium and the first said to be "on the air." Architecturally, it was even more mysterious than electrification: invisible radio waves flowed right through walls. Much as bright lights glorified big cities, so did the broadcast radius of a prominent radio station.[43] Much as electric lighting gave evenings at home many more possibilities, so did radio. Together, these technologies transformed the domestic sphere in ways that might seem primitive in comparison to today's enormous home theaters, but which were the greater departure from past practices.

Radio also helped annihilate distance.[44] It began the now-familiar practice of broadcast weather forecasts. Like other ambient media today, it was quickly appropriated for advertising (figure 10.5). Through its geography and its synchrony, radio created a unified audience in ways that print could not. It also diverted attention; as those who protest cell phone abuse today know only too well, a human voice talking to you out of thin air is harder to ignore than an embodied one engaged in conversation with others

Advertising by Radio Cannot Be Done; It Would Ruin the Radio Business, for Nobody Would Stand for It. Mr. McQuiston in this Article, Explains Why.

10.5 Editorial cartoon against early advertising by radio, National Radio News, August 1922 (earlyradiohistory.us).

present. For this, McLuhan famously called the medium "hot"—that is, not so ambient. Within a decade of its widespread adoption, radio had become notorious for its capacity to command attention, a power that was dramatically abused by Nazi and Soviet totalitarian regimes. Intellectual counterreactions formed a core of media-cultural theory that persists to this day. As first expressed by the Frankfurt School, broadcast by radio (and its successor, television) created undeniably spatial consequences in the configuration of cities for spectacle, distraction, and disinformation. Although this criticism began from the perspective of politics, it increasingly drew attention to media influences on the built environment, especially in the later age of runaway induced consumerism, epitomized by the "themeparking" of cities into predictable pleasure zones, where all transactions are commercial and documented, nothing of any consequence occurs unforeseen, and ever more aspects of human existence are monetized.

For a telling instance of the impact of information on environment, nothing quite rivals target market demographics. Mapping, mediation, and monetization increasingly combine. Although less astonishingly geometric than the Jeffersonian grid, these patterns, too, may be seen from airplanes, at least to the trained eye. The some sixty Claritas Prizm database categories, used by the target market demographics industry for nearly forty years, have reinforced differences among particular "lifestyle clusters" at the spatial resolution of the zip code, in a self-fulfilling feedback loop of residential and business site selection, to the point where the entire cultural landscape appears sorted.[45] As an agent of sprawl, these information technologies appear to have enforced economic and cultural stratification not only across the self-segregating social and political divisions for which the Internet age is so notorious, but also geographically.

Lifestyle segmentation, so evident on a regional scale, plays out on the personal scale by way of ambient media, especially ambient advertising. Consider screens in a room. Many if not most of the large flat panel screens visible outside the home belong to efforts at experience design to induce retail consumption. "Media" in the broadcasting and networking sense don't necessarily enter these efforts; just as often the video programming simply cycles through a locally installed loop about the experiences on offer right there. Tacit placement really counts. Advertising has, of course, long since transcended descriptions of products and services in favor of creating a lifestyle identity plus a general, indeed subliminally universal, awareness of the brand. Of all designers, advertisers most want their works to be ambient.

Because advertisers are so much more skilled (and numerous) than architects at connecting designed environments to sets of values and desires, there is a possibility that their contribution could increase, rather than erode, architectural awareness. Architect Anna Klingman has observed that branding can help communicate the "larger socioeconomic goals" of architecture and planning, and thus boost interest in built environments and "enhance the value that consumers place on spatial experiences."[46] Note the assumption that people are simply consumers. In counterreaction, if relentless sales promotions are not to crowd out moments for more genuine shifts of attention, new forms of information in environmental history need to focus on something more than wayshowing and branding.

American Space Reconsidered

In the United States, ideas of the commons tend to fall on deaf ears or, worse, to meet with ridicule. Is it a distinctly American trait to undervalue the space between things? Many of the cases and examples above would suggest so. Recent developments in reurbanization and urban informatics also suggest, however, that a watershed in environmental history might be occurring even in America.

"American space" was influentially explained by the motorcycle-trekking scholar of the vernacular landscape John Brinkerhoff Jackson, who explained how it took form most distinctly in the decades that followed the American Civil War. It is commonly said that America began the Civil War as many agrarian states but ended it as one industrial nation. New combinations of railroads, telecommunications, and commerce generated whole new landscapes, not only in the West but also in the reconstructing South. With these came new ways of orienting to the world. To Jackson, this era's shift from romantic to technological attitudes toward landscape was the single most formative phase in American environmental disposition.

Jackson opened his influential *American Space* (1972) with the coffee-table book, which better than any other information practice portrayed the formation of a new and enduring disposition toward technology-enabled production of space in the first post–Civil War decade. "It was the spectacle of America busting out of its historic confines and taking possession of untrammeled space that engrossed the journalists and compilers of commercial photo books."[47] This disposition was most evident in the smaller towns, where it produced the fenceless lawns and uniform rows of trees, especially in honor of the nation's 1876 centennial, that declared a "love of landscape"[48] Writing exactly a hundred years later, on village-green improvements as emblems of something much larger, Jackson observed "a tendency to subdue nature rather than cooperate with it, an unawareness of interaction instead of opposition between man and the world surrounding him."[49]

The age exalted second creation. Land was there to be developed, had little meaning unless developed, and could be developed in any manner its owner saw fit, often with total disregard for what others chose to do with theirs. In response to these disrespectful practices, Chief Seattle was said to have declared that "the land does not belong us; we belong to the land." A century and a half of "Don't tell me what to do with my land" has produced a chaos of building and discontinuity of landscape and, with it, an inherent disregard for what were once tacitly shared notions of commons.

As noted above, unprecedented wealth in the cities was tapped to remedy unprecedented squalor there. Not only waterworks and sewers but also gas lights, elevators, buses, and telephones gave rise to the central business district, the park-front mansion, and the streetcar suburb. Amid a progressive age, more creative work turned toward environmental criticism. Thus Jackson ended *American Space* with New York City, which, amid great vitality and a burgeoning population, began both to consider itself whole, as an environment, and to improve itself with monuments, with the great parks that did so much to establish the discipline of landscape architecture, and with new architectural approaches to the problems of its slums, approaches that could for the first time be understood as environmental.[50]

The ethos of second creation eventually came to an end, some say in 1916, when America encountered the horrors of the Great War, which shattered mechanistic notions of progress; others say in 1893, when the expansionist culture of Manifest Destiny ran out of land, and Frederick Jackson Turner declared the frontier closed. Yet the frontier mentality persists to this day, more than a century later, as if the world were a free good, and consuming it should be accounted as income and not capital depletion.

In sum, the attitudes of cultures toward their surroundings differ widely. Some value the world instrumentally; others, intrinsically.[51] Some emphasize things; others, connections. Some see the fish, so to speak, and others see the tank.[52] A culture with far more land than people, like nineteenth-century America, and one with far more people than land, like Japan, develop quite different philosophies.

Not even planetary change has shifted that bias very far. Despite a worldwide consensus on the need to become more responsible for surroundings, American thought, policy, and technology continue to emphasize isolated objects and events at the expense of their context. And that includes information technology, much of which is American in origin, if not best practices. Almost no other culture puts so many media feeds into quite so many aspects of life with quite so little regard for context. You can feel the cultural difference as you exit airport customs and enter this most distracted of nations.

10.	ENVIRONMENTAL HISTORY
Main idea:	Information deserves its own environmentalism
Counterargument:	Information leaves few traces
Key terms:	Environmental history of information, American space
What has changed:	More intrusive, inescapable media
Catalyst:	Bringing environmental sensibility to media
Related field:	Cultural landscape studies
Open debate:	Isn't information too ephemeral for environmental history?

Governing the Ambient 11

How much is enough? For information, often the best solution to too much is more: metadata, opinions, histories, filters, and background documents help lead the way. To restrict information would be unacceptable: the communications rights of individuals and communities must be inalienable, insuppressible, and not for sale. Yet among those rights might be ownership of your personal data, and a right to undisrupted attention. Thus when media become situated and persistent, profound challenges emerge in information ethics. Questions of stewardship, expression, privacy, pollution, and attention theft all intensify. To a generation of conscientious users, network philosophers, policy wonks, and street-level activists, these problems may long remain topics of policy and debate. The need for governance has become difficult to ignore. Whether through unanticipated liberation, ruthless privatization, or sheer volume, ubiquitous information technology now influences even the most everyday cultural acts. Under these conditions, it may be costly to neglect the role of augmented surroundings. For unlike the open Internet, an embodied system that people must inhabit imposes physical and experiential limits; more is not always better. Recall the consequences of those attitudes in mobility and housing. Now as shared physical spaces are flooded by media, corporations enclose cultural commons, and the dynamics of participatory networks shift to street level, what particular concerns arise with the ambient? Is there now a tangible information commons?

Recognizing Pollution

For one most obvious instance of these challenges, consider pollution. To ask whether information superabundance ever constitutes pollution seems especially problematic. Questions of intellectual and expressive freedom instantly arise. Seldom has any form of pollution been easy for a culture to acknowledge. Even chemical pollution of the air, however visible, was seldom discussed as such before the nineteenth century, didn't become a dominant cultural theme until the mid-twentieth century, and thus has a history that would mostly fit within one current human lifetime, if you date from, say, Rachel Carson's *Silent Spring* (1962) or the Clean Air Act (1970).[1]

Although today it is relatively easy to agree on what is wrong with dumping chemicals in the river, or that the right to dump anywhere may not be in the public good, filling the air with messages is another matter altogether. What is noise to one person may be vital communication to another, or to the same person on another day. To declare what pollutes is to censor, which is almost never a respectable act. Moreover, to claim the existence of information pollution is likely to add to it. How do you protest a problem whose solution is to say less? Yet freedom of speech is not the same as freedom to drown others out. The freedom of anyone to be heard requires some restrictions on the freedom of others to persist and to amplify. These are complex concerns.

In law, nuisance is often a matter of degree. An otherwise innocent act becomes an offense when done often enough, on a large enough scale, or in enough different places at once. You have the right to speak your mind, but not to amplify it to 100 decibels all night, nor to hang it on banners on a building that you do not own. So, too, a company has a right to promote its brand, but perhaps not in a manner visible twenty miles away, nor on the doors to City Hall.

If a loud party goes on into the wee hours, it is no surprise if the police show up. Noise pollution seems obvious enough. Many augmented reality projects now document it (figure 11.1). For a similar phenomenon, this inquiry has noted the "discovery" of light pollution. When is the focus of fixtures so poor, or the overall volume of artificial light so excessive, as to become unhealthy for all concerned? Somewhere, secret police subject detainees to unending brightness. Somewhere, vacationers pay top dollar for darkness, and a chance to do some stargazing.

Many towns now govern light pollution, with easily attainable advantages. By improving nighttime visibility while conserving power, this form of pollution control illustrates the possibility of doing more with less.[2] This begins from better design. Lighting experts say that you should see the effects and not the fixtures. The effect of visibility depends not so much on the volume of light as on the absence of glare, that is, on not having too much light coming from any one direction. As with

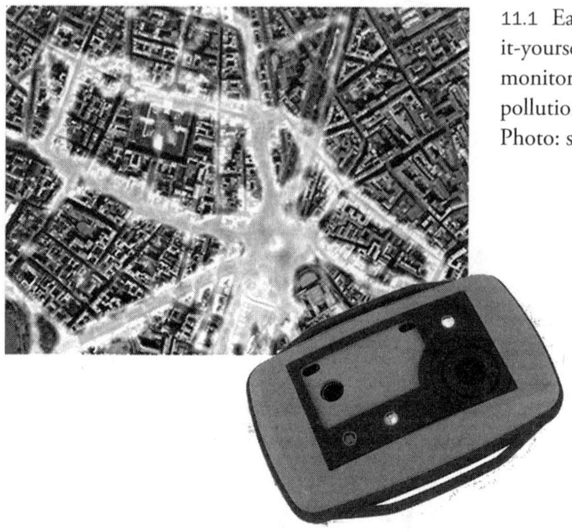

11.1 Early device for do-it-yourself environmental monitoring of noise pollution: Sensaris, 2009. Photo: sensaris.com.

conversations in a restaurant, speaking ever more loudly as background noise level increases definitely doesn't help; it only worsens the problem. Instead it is much more civil to find a pitch where you can talk under the background noise and keep your voice from carrying beyond your table. Similarly, light shouldn't trespass. By selecting, aiming, and masking light fixtures more effectively, everyone can see better while using less electricity. And with glare mostly gone, natural night vision has a chance to revive, and you can see some things with no lighting at all. You might also see the stars, and remember a larger (and planetary) place in the scheme of things.[3]

Often lit, now increasingly emitting light themselves, or just commanding the view in broad daylight, large outdoor signs demand similar consideration. Many towns have signage ordinances against full-motion billboards that distract drivers, for example, or against excessively tall sign pylons. You can clearly tell where such ordinances are being enforced when, on one side of a town line, there are no pylons at all and even the gas station logos hug the ground. The environmental graphics industry has learned these subtler ways in the last generation, much to the displeasure of kitsch enthusiasts nostalgic for early, now historic strips such as Burnet Road in Austin and U.S. Route 1 just north of Boston.

Back when competition among sign owners on the strips was wide open and automobile mania was new, an eminent art critic published a general environmental lament. Peter Blake's 1964 abundantly illustrated classic, *God's Own Junkyard*, was the most prominent work of the first wave of environmental recognition after *Silent Spring*. Actual junkyards were much more plentiful then, when cars lasted only a few years and material reuse was minimal. But to Blake, the real junkyards were the get-rich-quick commercial strips. "In destroying our landscape, we are destroying the future civilization of America,"[4] Blake warned, with what now seems uncanny accuracy. The signage depicted in *God's Own Junkyard* looks primitive and innocent by today's standards. Much less was known about marketing then. But, more obviously, there were also far fewer signage ordinances.

Many soon followed. Some previous exceptions first deserve mention. The federal court had ruled in 1917 that it was a valid exercise of Chicago's police power to require "consent of a majority of residents before a billboard could be placed in a residential neighborhood."[5] The landmark Highway Safety Act of 1936 had brought a degree of moderation to the era of serial roadside messages, like the famed Burma-Shave advertisements.[6] That early history of roadside signage has relevance today in debates over deadly texting while driving (figure 11.2). Only with the rise of general environmentalism in the 1960s did local signage ordinances become the norm in America, however. Following the publication of *God's Own Junkyard*, followed by the passage of Lady Bird Johnson's famous Highway Beautification Act of 1965, and eventually the publication of William Ewald and Daniel Mandelker's influential street graphics law manual in 1971,[7] a wave of governance swept across the nation. Since then, various cities and advertising lobbies have gone back and forth on the constitutionality of signage bans and separate laws for "commercial speech," as noted in the chapter on screens, with the case of Los Angeles's political exposé and moratorium on electronic billboards in 2009.[8]

Context is vital in signage law. The cultural geographer Wayne Franklin once observed: "You will not fare well as a literalist in the world of signage. Most signs require a fitting sense of context and, even more, a tactful regard for the slippage inherent in all social situations and most texts."[9] Thus, generally speaking, signage ordinances apply fewer restrictions to "on-site" signs; to avoid falling into censorship, they restrict the placement rather than the content of speech and in such a way that other outlets are always available.[10] And, as one sign industry organization explains, "many jurisdictions are amenable to a variance that would permit the renovation or retrofit of a building façade in order to enhance a district theme."[11] All told, signage law navigates delicate democratic issues for the greater good. Sign owners avoid the costs of runaway competition; localities avoid getting covered with references to someplace

11.2 Deadly distraction: an editorial cartoon by Herbert Johnson, *The Roadside Bulletin* (Vol. 2, No. 4), 1932.

else; citizens avoid having to look at quite so many eyesores; and yet nobody is silenced.

Achieving a more general communications commons seems much more difficult, however. Whereas if you don't filter the pathogens from the air, you could become chronically ill; if you don't filter offensive ads from your daily experience, you won't suffer any obvious health consequences. Parents at the supermarket take great care in selecting the foods they buy for their families. Yet, on returning home, they may leave the television on with no such alertness about the ingredients. Preoccupied with chemical pollution, they are oblivious to information pollution.

This inquiry began with a look at the meanings of "information," and the possibilities of overload or overconsumption. Although concerns about overload date back centuries, high anxiety over media that you cannot escape is far more recent. It began before smartphones, nevertheless. Fifteen years have passed since sociologist David Shenk gave an early warning: "Data smog gets in the way; it crowds out quiet moments, and obstructs much-needed contemplation. It spoils conversation, literature, and even entertainment. It thwarts skepticism, rendering us less sophisticated as consumers and citizens. It stresses us out."[12] As noted with respect to media facades, background music, and surveillance, to name a few, any future environmental history of information would do well to reconsider anxiety as an effect of information pollution, especially where media practices coexist and persist in shared sites of life. For, without a way to talk about these issues, there may not be much to do about them, or with them, except to filter your own feeds and tune the rest out.

Information Ethics

Superabundance amplifies the costs of poor information ethics. Many of these costs have become everyday realities: spam, scams, loss of privacy, pornography, identity theft, incivility, censorship, polarization, disinformation, piracy, spyware, shutdowns, denial of service. On one level, these ills seem to belong to more general problems in ethics, where culture cannot keep pace with technological change, and where public education in the fundamentals of political philosophy and of the social contract has fallen by the wayside. On another more obvious level, weak ethics belong to the frontier mentality of the Internet, where the advantages of new technology so outweigh the side effects that those increase as well. These regrettable ills and side effects might lie outside the scope of this inquiry if their becoming ambient did not so amplify them. So for the many experts in Internet culture, social science of media, environmental law, and ambient interface design who already have much to say about superabundance (as the solution to too much information really often is more information), this inquiry offers some questions (and not answers) on governing the ambient.

Many scholars credit cybernetics pioneer Norbert Wiener with the earliest expressions of computer ethics. Wiener foresaw an "automatic" age in which machines would have agency nearly equal to humans in the creation and governance of communications, which would play an ever more important role in society. He explained how the new computer age created gaps and ambiguities in customs and policies, and he issued clear warnings that profound moral choices lay ahead.[13]

Many long-term principles of information ethics come from the library sciences, which foster data stewardship, especially in seeking to keep the most accurate information most readily available: don't introduce bad data; don't lose data; don't forget where data came from; and don't restrict access to data. Data integrity has become a central ethical challenge, made all the more difficult now that so many amateurs across the globe are producing and sharing data. Thus information ethics of the early web era focused largely on the acts of individual users, and the automatic agents they released.

Today, however, information ethics just as often involves notions of commons and threats of enclosure. For example in the upkeep of commons, and in contrast to copyright, there exists a notion of "copyduty." In enclosure, culturally shared names of long-existing places become property of corporations. Information ethics is highly participatory and the stuff of everyday Internet policy debate. Cory Doctorow, a luminary in this debate, has explained how the ethics of copyright have been distorted by the "agreements" you have to accept even for personal use of media artifacts. As originally conceived, and as usually practiced, copyright was not intended that way: "The realpolitik of unauthorized use is that users are not required to secure permission for uses that the rights holder will never discover."[14]

Among other gaps in information ethics, not enough has been done to make "attention rights" into an everyday ethical theme. Filling this gap begins by protesting how intrusions have diversified. Thus, as Internet strategist Tom Hayes has observed, commercial interruptions were predictably part of the bargain in 30-minute television programming, but they operate differently and more annoyingly in the more selective, real-time experience of the Internet. "As marketers and advertisers hungrily explore ways to monetize online attention, they face mounting challenges. Consumers have migrated online precisely because they want more control over the media they consume. The old bargain [from broadcast television]—content for attention—is broken."[15] Precisely because so much more now occurs out of context, badly placed media cause upheaval in the economics of attention. Might violations of these rights be one main source for widespread perceptions of overload? Hayes foresees an attention rights movement. Among the principles in his seven-point manifesto: "I am the sole owner of my attention;" . . . "I own my click stream and all other representations of my attention;". . . "Attention theft is a crime."[16]

Much as a citizen has a right to be heard, and not to be silenced or drowned out by more powerful players, so also a citizen has a right to attend, and just as importantly to choose not to. This suggests a right of ownership of attention. A society has a duty to prevent thefts of that. Normally, this begins in civility. Respecting the mental life of others, especially amid the conduct of civic life, civility upholds the right to reflect. Wherever speech is a right, civility moderates how that right is exercised. Wherever self-interest prevails, civility needs careful upkeep, especially in this age of disembodied media. Legal scholar and novelist Stephen Carter has characterized civility as "the sum of the many sacrifices we are called [upon] to make for the sake of living together."[17] Community life requires often placing the common good above immediate self-interest, even when associating with strangers. Technology sometimes undermines this principle. For example, as Carter has emphasized, the automobile has eroded the need to get along together: going everywhere alone in a cocoon of comfort has created a sense that convenience is everything, and that other people are mostly in the way.[18] Likewise, perpetual messages supplant going to see someone, and scripted experiences, such as at the retail point of sale, studiously suppress spontaneous conversation.[19]

The civility of individual information acts shapes and is shaped by their informational environments, whether technological, societal, institutional, or biological. Information ethicists increasingly put these environments first.[20] Environments have intrinsic value. The very existence of cultural artifacts has value, and the environments where these are aggregated deserve stewardship.

As in a general ethics, information ethics attributes value to the very being of something, extending this value to cultural productions such as works of art and even to personal data sharing. It assigns a moral value to the preservation and protection of informational entities. Thus it is not just for the subject or agent who obtains, reads, copies, or sends information, but also for the information entities themselves. As explained by epistemologist Luciano Floridi, from whom the above summary is paraphrased, information ethics applies both to individual resources and entire systems, to protect them from what he calls "entropy."[21] Moreover, it assigns moral agency to the organizations, artifices, and distributed actions that operate over networks. This makes it a vital component of a general ethics of twenty-first-century life. All operators on information must understand and preserve their environments. Key to Floridi's view of informational ethics, this responsibility extends from micro to macro scales—from individual acts of obtaining or disseminating information to what Floridi has called the "infosphere." This makes information environmental.

Networked Commons

Like attention, everyone knows what a commons is—or so they think—yet misconceptions abound. First off, nostalgia enters the picture. To Anglo-Americans, for instance, a commons was a simpler community arrangement disrupted by industrialization; in preindustrial England, it was a meadow for grazing livestock; in early New England towns, it became a place to assemble for democratic process. Today, in its widespread misuse in naming suburban condo subdivisions, the word carries nostalgia for such origins. Even in its more thoughtful senses, as for data sharing or communities of practice, the word *commons* gets dismissed as romanticism. For as everyone knows, the tragedy of the commons is that, through pursuit of enlightened self-interest, individuals collectively exceed the total carrying capacity of the commons and deplete its resources. Worse, any mention of "commons" in a capitalist society, where the market is expected to solve all problems, might easily be dismissed as naïve socialism.

Fortunately, the rise of a more networked age has established that commons are not socialist regimes, nor indeed states or markets at all, but are instead necessary complements to both states and markets. Perhaps no better explanation of this exists than the recent contribution by Lewis Hyde, *Common as Air*. By now, most scholars know that the oft-cited tragedy of the commons doesn't describe a commons at all. Indeed, "Garrett Hardin has indicated that his original essay should have been titled 'Tragedy of the Unmanaged Commons,'" Hyde's work explains, "though better still might be 'The Tragedy of Unmanaged, Laissez-Faire, Common-Pool Resources with Easy Access for Noncommunicating Self-Interested Individuals.'"[22] This encapsulates more recent thinking that a commons is not an open rivalry for resources so much as a set of upkeep measures among a managed network of participants seeking both individual and joint benefits. *Common as Air* enumerates many sets of management measures, past and present, with emphasis on *stints*, that is, frugal allotments. As a verb, to stint is to use less than you might, "only enough, with as good left for others."[23] As a noun, a stint is a voluntary restraint—something societies in decline generally lack. For example, traditional markets had stints, such as first access for locals, access only on designated days, and so on. Indeed each sphere of life had its own stints, often against the others. To govern all aspects of life on the principles of any one sphere (whether market, church, or state) was an invitation to tyranny.[24] Today of course the tyrant is the market itself; and what this tyrant forcibly (through lawsuits) seeks to obtain is ownership of knowledge, even of attention.

Thus a commons exists as rights of action, and in the networked practices that stint them. This helps resolve small conflicts over uses before they became large and contentious. Acceptable boundaries, ways of monitoring, and scales of penalties are all negotiable. Whether explicitly encoded, implicitly practical, or both, such a governance enacts an ethics of situated information. It arises from patterns of use, often in what today are called microtransactions. For it is a fundamental fact that information technology improves many such patterns, and that these social patterns, and not just the resource they manage, are the commons. The late Nobel laureate Elinor Ostrom may have been the first to summarize these principles for an age of networked, resource-managing organizations. Most notably, from Ostrom's summary list: "Rules in use are well matched to needs and customs; . . . Individuals affected by these rules can usually participate in modifying these rules; . . . A system for self-governing members' behavior has been established; . . . Community members have access to low cost conflict resolution mechanisms."[25]

"Neither the market nor the state," Ostrom's oft-cited slogan, helps explain how networked organizations compensate for the oversights and shortcomings of those more distant spheres. To activist entrepreneur Paul Hawken, networked organizations are nothing less than the human component of a planetary immune system. To Hawken, the members of these organizations "were typically working on the most salient issues of our day. . . . They came from nonprofit and nongovernmental world, also known as civil society. . . . I initially estimated over 100,000 organizations working on ecological sustainability and social justice. I now believe there are more than a million."[26]

Wherever arrangements arise to complement what markets can and cannot value—especially to manage upkeep, access, and experience of resources by means of networked noncommercial transactions—there the idea of a commons takes new form. Today, there are many kinds of commons (figure 11.3). "The language of the commons," activist David Bollier has asserted, "serves a valuable purpose. It provides a coherent model for bringing economic, social, and ethical concerns into greater alignment."[27] These concerns exist because some resources are inalienable, such as attention. Otherwise, markets tend to diminish regard for anything that cannot be bought and sold.

Many more organizations have come to recognize that treating the drawdown of human, cultural, or natural capital as income constitutes a serious problem, and have used information networks to uphold the intrinsic, and sometimes also highly instrumental, value of these resources. The Internet became such an environment in the 1990s, amid the explosive rise of the World Wide Web. As Ostrom and coauthor Charlotte Hess later reflected:

> The "information commons" movement emerged with striking suddenness. Before 1995, few thinkers saw the connection. It was around that time that we began to see new usage of the concept "commons." There appears to have been a spontaneous explosion of "ah ha" moments when multiple users on the internet sat up, probably in frustration, and said,

> "Hey! This is a shared resource!" People started to notice behaviors and conditions on the web—congestion, free riding, conflict, overuse, and "pollution"—that had long been identified with other types of commons.[28]

Like so much else at the time, this particular notion of commons was conflated with cyberspace, virtual reality, community networks, and digital civics. This was before talk of Web 2.0 or the rise of social software, especially for resource sharing. It concerned technological prospects for what was then a young discipline of information architecture—not naive notions of immersive visual data navigation as in the novels popular at the time, but ambitious agendas in ontology, findability, and other such forms of server-side stewardship that computer scientists-turned-service-designers thought about. (They still do, but with less venture capital being thrown at them.) This early phase of information commons anticipated something other than today's corporate creed of the Cloud, or today's nonmarket realities of social production. From an age before broadband, it was less about traffic, and more about stewardship. After all, the Internet began among researchers who used it partly as a data commons. This raises two obvious questions: Are the peer-reviewed papers of an academic society a knowledge commons? Are the petabytes of data streaming around universities a commons?

In many ways, the event that coalesced the idea was the foundation of Creative Commons in 2001, which made it clear that an information commons was not cyberspace writ large, but an intellectual property regime. This was surely an act of recognition: clearly the web was a cornucopia resource, that is, one where each user increased rather than depleted the resources shared. For, as any YouTube contributor now knows, much

	Organization	*Depletion*	*Participation*
Not a commons	Free enterprise	Free riding ("tragedy of the commons")	Defense by owners
Physical commons	Designated area	Sustainable if well managed	Conflict resolution by members
Information commons	Networked access structure	Cornucopia: more use makes more of it	Contributions by millions
Ambient commons?	All the above?	Total cultural white noise?	Curated by inhabitants?

11.3 Kinds of commons.

scarcity is purely artificial. The net was more a "comedy of the commons."[29] Here more use made the resource better, and copy-duty was to share and share alike.

Instead, ten years later, copyright law has become a main battleground of Internet policy. "If democratic practice (not to mention creativity) depends on plural speech and plural listening," lamented Hyde, "we should be reluctant to give any modern form of Negative Voice [similar to a king's censorship or veto] a presence in the public sphere. But of course we have."[30] For instance, there appear to be too few stints on media companies searching private hard disks for even the smallest copyright offenses, and then threatening to deny access, simply on the grounds of accusation, for what by now is a common carrier necessity of Internet service. (As America's ill-conceived Stop Online Piracy Act would have done, had it passed in 2011.) The absurd extent of unstinted privatization can be seen in physical space as well. Thus, in one review of a pop art exhibit in London's National Portrait Gallery, with work of original mashup appropriation artists such as Andy Warhol on display, Cory Doctorow found that he was not even allowed to photograph the "No Photography" sign. "If true, presumably the same rules would prevent anyone from taking any pictures in any public place—unless you could somehow contrive to get a shot of Leicester Square without any writing, logos, architectural facades, or images in it. Otherwise, I doubt even Warhol could have gotten away with it."[31]

In other words, early conceptions of information technology as a cornucopia commons arose before widespread recognition of how networks help turn ideas into commodities. The problem had been recognized as early as the rise of the personal computer. In 1983, Ivan Illich cautioned that "computers are doing to communication what fences did to pastures and cars did to streets."[32] By the dawn of the new millennium, as corporations ever more aggressively privatized knowledge and cultural identity, this effect became known as the "second enclosure movement." According to James Boyle, who is credited with that expression, "once again, things that were formerly thought of as either common property or uncommodifiable are being covered with new, or newly extended, property rights."[33] Genetics research provides the most salient examples, but enclosure also happens in everyday life: the naming of ballparks, the appearance of particular retail districts, traditional stories made into Hollywood movies, and even some words. According to Hyde, in mass consumer culture, "the young are taught a language that is not theirs to own."[34] There must be a right for a community to own its speech. There should be ways to know without being instructed. There should be rights to conversations without covert surveillance, to hold third grade class without commercial messages, to assemble in the park with no declared purpose, to use cash as legal tender, and to see the night sky.

Tangible Information Commons

What would make a good set of cultural stints at street level (figure 11.4)? The noise, light, and signage ordinances noted above only begin to address the challenge presented by embodied media. Thus they say nothing about whether walking past a store might cause a popup ad on your phone, or whether it would be wrong for that ad to occur in audio, either as an interruption to a stream on a device, or as a targeted beam in the space of the sidewalk. Signage law has few restrictions on how a physical site might search your phone for history data, like websites do with cookies on your hard disk, so as to decide which content to display on an outdoor screen as you walk by. But then unlike a message on your laptop or smartphone, not only you but also incidental bystanders would see an embodied media message.

In the design and governance of shared built space, information ethics can improve environmental experience. By giving intrinsic information its chance and by forming digital information into more appropriate textures and resolutions, design practices make some spaces more usable, while also letting the world speak for itself. Although some people surround themselves in an ever greater array of media, and others simply seek ways to

Data formation
When data are everywhere, design their form and resolution, for scale, texture, figuration, and duration.

Embedding
Integrate architectural features and assemblies; don't cover them. Let intrinsic structure show, especially when powered down.

Participation
Anyone who has to live or work in the presence of a large media surface should have a say on what appears there.

Thematic layering
Use augmentation to focus and filter local data increases. Reduce the amount of data perceived by those who don't want it.

Slowness
Don't thoughtlessly use the time frames of cinema, television, video, or flash. Some display might change imperceptibly, over hours, days, or longer.

Camouflage
Use the principle of least effective difference. Get attention by resembling what is receiving attention anyway.

Locality
As in signage law, restrict references to subject matter that is off site, and enhance references to what is right here.

11.4 Seven stints for an ambient commons.

unplug, all seem to manage their attention in some new way. Some of these choices involve personal will and discipline. Other choices involve law and civility. But together, these choices lead the way out of a growing sense of overload to more reflective lives amid more richly layered latent possibilities for attention. In other words, there may somehow emerge some prospect for governing the ambient.

The considered life requires a balance between messages and things, between mediated and unmediated experience. Citizens have a right to engage one another and the built world they inhabit in ways that are unmediated, uninstructed, unscripted, and undocumented. Under an ethics of preserving and protecting existing information environments, there should also be a right to preserve the subtle high resolution of the intrinsic structure of the local world, to protect it from being covered over with the crude low resolution of one-size-fits-all media productions. Does having more ambient information make you notice the world more, or less? Can mediation help you tune in to where you are? Or does it just lower the resolution of life?

Today, ambient information media become more difficult to escape. They channel more kinds of communications into shared physical contexts that, like city parks, come with expectations of being commons. They also take on aspects of commons as they make the city more usable. So far the ambient commons is just an idea, and hardly a usual one. Yet in its potential for becoming a commons, the ambient is a very rich cultural challenge, and no simple meadow.

11.	GOVERNING THE AMBIENT
Main idea:	Governing ambient information, as if it were a commons
Counterargument:	There is no commons
Key terms:	Nonchemical pollution, second enclosure movement
What has changed:	Ubiquitous media in shared spaces
Catalyst:	Threats to cultural commons
Related field:	Information ethics
Open debate:	Is information environmentalism even possible?

Peak Distraction 12

Overload and distraction have always existed. The question is whether they have become more ubiquitous and subtle. The natural workings of attention do seem less well adapted for a world full of superabundant, artificial, often cognitively engineered stimulus. These conditions suggest a new stage in the history of information, one quite different than even ten years ago, much less centuries past.

Under these changing circumstances, your attention practices may feel stressed in new ways. They may require some new kinds of mental health breaks. Just filter better, the sages tell you, and use the latest technologies to do so.[1] But you may sense a limit to what you can filter, especially in physical space, and a steady encroachment of potential distractions there. The audio volume at the grocery store is set slightly higher than last year. Advertisers have found yet more places to insert their messages. Everywhere there are ever more public safety measures. (How did anyone survive all those years without them?) Altogether there are many more autonomous annoyances—both online and at street level.

Today's increasingly ambient information practices demand a different kind of sensibility to surroundings. Henceforth, may you seek, inhabit, and maintain surroundings that are less thoughtlessly layered in media, and more discriminately curated for use. There, as quiet fascination becomes more of a habit, may a virtuous circle arise (figure 12.1). In what may well internally be a process of neuroplasticity, may new fascinations and sensibilities make you less distractible in the first place, or somehow less prone to overconsumption. Whatever the benefits of leading a well-focused life,[2] as countless productivity and spiritual purpose guides would gladly help you do, there also benefits in dwelling well, through good design, and in taking notice of where you are. It is for broadening that particular circle of sensibilities that this inquiry has occurred.

In hindsight, from the perspective gained by this shift of sensibilities, it might be possible to recognize "peak distraction."[3] It should be possible to look back in wonder at just how indiscriminate ambient information practices had been. For example, today's historical perspective makes mid-twentieth-century obsession with cars seem indiscriminate. Why did they try to do everything in cars? What was it like to live with so little walking? Who let them reshape the built world just to keep the cars moving? Couldn't they foresee any of the eventual costs?

Hindsight might reveal beliefs that ubiquitous feeds had no cost. People assumed having more messages would always cause more belonging. They tolerated in-your-face media, assumed needs for perpetual entertainment, lazily consumed whatever was playing, and generally tuned out of shared space. But then, one by

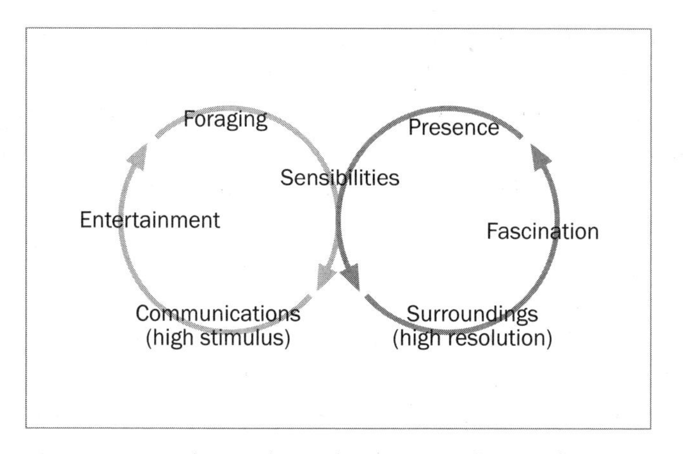

12.1 Virtuous circle: cumulative adaption (neuroplasticity) from noticing surroundings too.

one, and town by town, they began to exclaim "Enough!" As they proceeded to alter (but not eliminate) so many ambient information practices, distraction peaked, at least for some of them.

To declare peak distraction is not just to take more mental health breaks. It is not a wish to go back to some earlier way of life, nor just to tune out further from the present one. It must include some participatory new sensibility to surroundings. If enough people seek this in the spaces they share, there may yet be an ambient commons.

Synopsis and Retrospect

This has been an inquiry and not a scientific proof, nor a journalistic exposition. It has employed a superabundant range of source materials, with the advantage of university research library tools and databases. It has taken a longer time perspective than technologists usually care to do. It has assumed continuing need for long-form arguments and secondary research. Thus it has done some of what a scientist, historian, journalist, designer, or blogger might do, and with an aim to consolidate and not just aggregate, as if the aim is a longer but easier trail through so many ideas.

One more disclaimer: this solitary inquiry has not dwelt on social media, which are changing all too quickly, and whose instant short format seems quite the opposite of print. It admits how, to ever more people, the social network has become the locus of knowledge, or even being; and that to many of them, the idea of wanting to be alone with one's thoughts might seem baffling. This inquiry accepts how that might be true for accomplished researchers and not just teen social life. More ambitious social (and academic) networks do gather, broker, maintain, and debate knowledge in ways that a solitary scholar (or reader) could never do. This all seems quite Jamesian: knowledge becomes more dynamic, approximate, statistical, and provisional. In this sense, the Internet shakes the university to its core; the two are now breeding a new heir. What all this does for individual knowledge and attention will take the best philosophers of the day (both solitary and networked) to investigate.

To summarize the present argument for an environmental history of information as presented in previous chapters. New media have been creating a middle ground for "urban markup." Neither official signposting nor criminal graffiti nor business as usual in advertising, urban markup invites new curation, civility, and restraint—but by whom?

An explosion of contexts and formats for situated images—screens everywhere—has been altering visual experience, at least at street level. There are so many screens everywhere that there is little impulse to look at any one of them, much less through them, as if apertures onto someplace else. Instead the experience becomes postcinematic, often architectural, and increasingly about reading the built environment.[4]

A more truly ambient experience of built form comes from the engineering of atmospheric comfort. As energy imperatives and smart green building technologies combine to move architecture beyond the sealed glass box, atmosphere ceases to be so uniform, and embodied cognition suggests new relations of commodity, firmness, and delight.

The experience of the city seems even more enmeshed in digital media practices on a still larger scale, especially wherever the smartphone has become the main way of getting on with business. Sensemaking unfolds not only on the scale of a single work group or specialized site, where habitual actions lead into a belated understanding of a situation, but also on the scale of daily life across the metropolis.

Even the simplest look at environmental history suggests the importance of recognizing the stages in the relations of nature and artifice. Different people, disciplines, or cultures sometimes coincidentally identify some new overarching phenomenon, such as pollution. Histories of lighting and signage have a place in this, as do broadcast media, although the rise of situated information technology raises major new themes. But is information environmentalism even possible, much less desirable?

The Internet has plenty of commons advocacy, yet is mostly libertarian (as well it should be). The idea of a commons plays differently for information cornucopias than it does for depletable, pooled physical resources. But which of these governs the accumulation of situated media at street level? This suggests a tangible information commons, or at least new practices in information ethics.

To understand such turns in the environmental history of information, it helps to know more about the workings of attention.[5] For likewise there have been dramatic shifts in the discipline of cognitive science. Context and embodiment now appear to count for much more than was believed a generation ago, when many of the foundational principles of digital media and computer-like models of mind were first established. New stages in the histories of environment and information demand new insights into attention. The principles of embodied cognition that have been explored here as the workings of attention do suggest how persistent and inhabitable phenomena should mean more than they have, both individually and culturally.

The cognitive role of architecture is to serve as banks for the rivers of data and communications, to create sites, objects, and physical resource interfaces for those electronic flows to be about. At the same time, architecture provides habitual and specialized contexts by which to make sense of activities. And, where possible, architecture furnishes rich, persistent, attention-restoring detail in which to take occasional refuge from the rivers of data.

In the meeting of architecture and interface, attention and situated technology, or cognition and environment, embodiment has long been a unifying theme. By working within that intellectual tradition, perhaps this inquiry has shed useful new light on the nature of attention to surroundings. That seems largely a matter of habits and practices. Although natural limits exist, as to multitasking, attention practices are adaptable in powerful ways that are better understood than they were even a decade ago. Neuroplasticity exists, makes use of embodied cognition, and is worth knowing and debating more often. The brain is neither fixed for life, nor freely malleable into just anything, but instead is patterned, emergent, and recombinant. It acquires pathways, especially from habit, especially with context, and tends to use elements of those for other incidental purposes along the way.

You have seen how it helps to discern different kinds of information itself. Instead of the old technocentric definition of information as sent symbols, it helps to take the epistemologists' definition as "true semantic content." It also helps to qualify the latter in two further ways. First, content may be something you do and not something sent to you. Second, not all information is semantic; you may also be informed by the intrinsic structure of things.

As a noun, *ambient* conflates many such considerations into a word sticky enough to represent a new sensibility. No longer just a luxury, the ambient has become subject matter for creative work by anyone who occupies or augments shared physical spaces. In an age of superabundance, it is important to tune the balance (as the philosopher Albert Borgmann so well put it) among information about, for, and as the world. In a world after peak distraction, the design, curation, and experience of the ambient all begin to improve.

Sensibility, in the End

Any survival guide to the twenty-first century includes instructions for the educated person on what to do in the event of data smog, hive minds, engineered distractions, attention theft, compulsive connectivity, coercive connectivity, and the like. Many of these instructions involve a casual fascination with context—habits of awareness that work almost anywhere, enough so that you can be perfectly content sitting stuck in traffic without music playing, and without messages to check.

On the whole, well-being involves attentional balance. It takes continual reeducation of the brain. Sensibility means cultivated attentional preferences. It means a refined receptivity, such as a musician's ear, an athlete's sense of position, or a tongue-tied traveler's eye for body language. It also means a more engaged disposition than tuning out. To cultivate the selective sensitivities now necessary in an age of embodied information, you need to take notice of more, not less. The blasé detachment that may have worked best against industrial and broadcast events now just leads to much more solipsistic overload. Instead, your better defense is in the details of everyday life. As with food, so with information, it helps to value the slow and the local.

In the end, this is an argument for good design. Design sensibilities must become much more widely appreciated for how they configure larger attentional sensibilities. Now comes the time to reconfigure many built contexts for less waste on a changing planet. Information superabundance stands to be a major factor in that. Better design sensibilities toward ever more situated technologies now seem vital. But, for a start, just take stock of attention. Take back your attention, again and again, give more and more of it to your surroundings, and help take care how those now seem to be filling with new media.

In an age of distraction engineering, you have no choice but to manage your attention more mindfully. To do so requires more knowledge of how the brain works, but also of how the built world works. So when choosing the focused life, don't forget to choose where. This isn't just about stopping to smell the roses, although that sometimes helps. A new mindfulness to context becomes no mere luxury when the world becomes augmented, and the ambient takes form.

12.	PEAK DISTRACTION
Main idea:	Personal sensibility includes surroundings
Counterargument:	Anyone can already filter however they please
Key terms:	Peak distraction
What has changed:	Mindfulness about overconsumption
Catalyst:	Neuroplasticity
Related field:	Any reflective practice
Open debate:	How much can anyone filter?

Epilogue: Silent Commons

Monastic silence: instructions for tourists at the high Benedictine abbey of Fleury, France.

You should have the right for your environment to remain silent. Without that silence, you may get fewer chances to speak your mind, to hear yourself think, to find empathy for others, and somehow just to be. "Silence, according to western and eastern tradition alike, is necessary for the emergence of persons," Ivan Illich once cautioned. In a 1983 speech called "Silence Is a Commons," Illich offered a parable on the encroachment of media. This was long before mobile phones, sensor fields, or media facades:

> On the same boat on which I arrived in 1926 [to the Dalmatian coast, as a newborn], the first loudspeaker was landed on the island. Few people there had ever heard of such a thing. Up to that day, all men and women had spoken with more or less equally powerful voices. Henceforth this would change. Henceforth the access to the microphone would determine whose voice shall be magnified. Silence now ceased to be in the commons; it became a resource for which loudspeakers compete. Language itself was transformed thereby from a local commons into a national resource for communication. As enclosure by the lords increased national productivity by denying the individual peasant [the right] to keep a few sheep, so the encroachment of the loudspeaker has destroyed that silence which so far had given each man and woman his or her proper and equal voice. Unless you have access to a loudspeaker, you now are silenced.[1]

The loudspeaker had violated some tacit expectations. In reference to the traditional cultures of Japan, where he was speaking, Illich observed:

> People called commons those parts of the environment for which customary law exacted specific forms of community respect. People called commons that part of the environment which lay beyond their own thresholds and outside of their own possessions, to which, however, they had recognized claims of usage, not to produce commodities but to provide for the subsistence of their households. The customary law which humanized the environment by establishing the commons was usually unwritten.[2]

Silence remains in such trust. Silence is not the totality of having been silenced, but the ambient reflectivity from which anyone may speak. Silence remains necessary for individual and especially cultural sanity. This is why modern cities enforce noise ordinances. You have a right to free speech, but not to amplify it from the rooftops all night, as if the air were an inconsequential void. Many cities, such as Chicago, will now impound "boom cars," whose hood-mounted speakers thump too loudly.

Of course, the boisterous diversity of sounds is the very stuff of city life. To move about is to make some kind of noise. Life seeks the clamor of the marketplace, the exuberance of festivals, the roar of the crowd at the game. In many walks of life, to sit silent is complicity, submission, or alienation. It is literally to be dumb. Yet little is more dumbfounding than to lose control of your attention. Wherever advertising, entertainment, newscasting, or total public safety instruction dominates all communication, there the art of conversation languishes. There the ears might shut down. Perhaps texting has become so popular because it requires no listening.

In the millennial year 2000, British novelist Sara Maitland chucked it all and moved out onto the moors, where she soon "fell in love with the silence." In 2008, she published *A Book of Silence* as a chronicle of her attention restoration. "All silence is just waiting to be broken!" a close friend had chided her at the outset. "Silence is not for everyone," Maitland replied. Although almost everyone claims to want a little peace and quiet, most avoid prolonged silence, as if it were "dangerous to our mental health, a threat to our civil liberties."[3] If some find palpable silence just as oppressive as excessive noise, perhaps they do so from a very different set of beliefs.

First, some instinct rightly exists for the inherent boisterousness of life; people want to hear one another living. Second, at some acquired cultural level, people rightly fear oppressive authorities and institutions, as if silence means subjugation. This seems so whether one is silenced by force, as with Nazi loudspeakers, or by one's own submission, as in monastic self-abnegation. Indulgent noise dispels such threatening associations. Third, today's social technologies make the silence of solitude suspect. Although perhaps any tribal village culture would likewise have taboos against silent loners, today's social media reify that bias, and make communication unprecedentedly frequent. Quiet time barely exists; indeed the word *time* increasingly implies a point and less often an interval. Fourth, although Maitland didn't dwell on this, perhaps some people would just as soon not be left alone with their thoughts, as these days those tend toward doom and gloom. As a result, most people "choose to have incessant sound pumping into their environment,"[4] as Maitland put it. Still, it wasn't the tunes from the ceilings that led her to plot her escape, but an "encounter with positive silence."[5]

Like the air on which it is borne, silence is not just an absence. As a presence, it has always attracted the life divine. The mere seeking of this presence, whether as an unsullied planetary Nature, a personified God, an unnamable immanence, or just the sound of your heart and your nerves, admits that something is missing in a world of endless stimulation. To the acclaimed Estonian composer Arvo Pärt, silence was a single note: "I am alone with Silence. I have discovered that it is enough when a single note is beautifully played. Silence comforts me."[6] To practice such presence is to uphold how you can become informed without transmissions, codes, or things. In an often-told anecdote about Mother Teresa, her followers wanted to know what she asked God in prayer. "Nothing, I just listen," she replied. "And what does He tell you?" they inquired. "Nothing, He just listens."[7]

The musician John Cage, who was no saint, offered what would become his most famous work, "4'33"," four minutes and thirty-three seconds during which the performer was to sit at a piano in silence. As the mathematician Douglas Hofstadter once observed, there is many a pub where you might wish that "4'33"" were on the jukebox.[8] Cage explained his desire to rethink listening. He wanted to reawaken the ear to the patterns of sound in the world. He took those as music—"poetry to his ears."[9] This purposeful acceptance of extraneous sound as music aptly reflected the increasing hubbub of twentieth-century life, where the experience of silence was quite rare. Here was an early recognition of attention restoration.

Total silence would be rare in any case. Writers, most of whom seek silence daily so they can hear the sentences they are making, sometimes write of the quest for total silence. Some such as Aldous Huxley sought it by floating in sensory isolation chambers. Wondering if there was any silence to be had in New York, novelist George Michelsen Foy eventually looked elsewhere: alongside monks, in the experience of astronauts, or deep in mines.[10] But even there, your heart still beats, and depending on the life you have led, your ears might never stop ringing. Like national security, quiet is relative; totality is a misconception. Instead, you must seek small intervals of somewhat less noise, as when stadium crowds pay their respects with a "moment of silence."

Silent fascination can restore attention.[11] Running on effortless perception of natural data, fascination exercises neural resources other than those fatigued by excess stimulation. Thus, to recover some attention, you might spend a hour with the sounds of a waterfall. There you might wave hello to young ear-pod-equipped hikers who plow by without breaking stride; but if you suggest they could be missing something, they might simply roll their eyes. Maybe they like soundtracks with their waterfalls. Maybe seeing is enough. Besides, the woods are increasingly filled with the sound of off-road vehicles. The park experience, once vaunted as "America's best idea,"[12] involves ever more noise from vehicles driving by, flying over, and churning or splashing past. Garret Keizer, the keeper of noisestories.com, has compiled an impressive portrait of such annoyances in the book *The Unwanted Noise of Everything We Want.* To Keizer, as to almost anyone who seeks some occasional quiet outdoors, "recreation" loses something when motorized, for bystanders and participants alike. Yet the heavily equipped "sportsman," atop his costly off-road rig, quite likely hauled by a rugged, eagle-emblazoned truck, may quickly dismiss any frugal, silence-seeking hikers who dare to protest as "elitist."[13]

Restoration also begins from stillness. Even in small, untrained, and far-from-daily sessions, meditation conditions the ability to identify and let go of thoughts, messages, and images as ephemeral features of the mind's more fundamental being. Noises may nevertheless penetrate. Even a seasoned teacher still hears a siren go by outside on the street. But then that noise washes away. Much else soon follows. Desires may vanish, such as for recorded music. It becomes possible to be comfortable sitting in traffic without any audio in your car. Abilities may meanwhile arise, such as to tune out the autonomous annoyances of the augmented city, to tolerate nearby cell phone conversations, or to marvel at the exuberant expression of your neighbor's mechanized assault on his yard. But, most of all, some ability to discern among kinds of silences and noises begins to take hold. For instance, a difference becomes clear between the kind of silence that discourages speech and the kind that offers enough quiet for conversation. Likewise, you might note a difference between the noise of life as it always has been and the noise of recent new forms of overconsumption.

Much noise does correlate with consumption, as Keizer has explained, and with disregard for anything that can't be bought and sold. More precisely, producing noise does so; whereas having to live with noise correlates with poverty, or worse.[14] In the industrial age, the poor had to live by the railroad tracks; but the affluent moved to the suburbs, where they could run their lawn mowers in peace and quiet. (Of course, today fewer of the affluent mow their own lawns; the professional crews that now provide that service employ huge tractors, weed whackers, and squadrons of leaf blowers; any sense of reflection once part of keeping a lawn seems to have vanished.)[15]

So despite how unimportant it may seem in comparison to the major issues of these times, noise deserves more reaction. As Keizer has put it, there is parable of disregard for the planet in how so many of those who produce noise so easily dismiss whatever they are drowning out. Who needs conversation? Who cares about the fools who claim to care about birds? Keizer's main concern is arrogance: "In the end, after all the physicists, musicologists, and social theorists have had their say, there are only two kinds of human noise in the world: the noise that says, 'The world is mine,' and the noise that says, 'It's my world too.'"[16] Whereas some noise necessarily and incidentally emerges as a byproduct of living, working, and moving about, Keizer hears a new kind of deliberate and unnecessary noise taking over. It is the kind of noise that continues long after those who started it have gone home.[17] To protest this secondary and often subjective form of pollution might seem churlish and even elitist. But in deft response to these usual objections, Keizer shows how mindfulness to noise abatement connects with awareness of environment. He shows how so much noise comes from the assertion of self-importance over respect for environment, apparent not only in obvious offenses such as jet skis, but also in everybody-does-it acts like taking a commercial airline flight. Thus the cultural shifts now necessary for survival might just begin with learning to live more quietly.[18]

A quieter life needn't feel sad or solitary; quite the contrary. It involves participation ("my world too") in the temporary making of unmediated human noise by all, in the upkeep of physical and cultural commons, and in the curation of the media that inevitably infuse the world. A quieter life takes more notice of the world, and uses technology more for curiosity and less for conquest. It finds comfort and restoration in unmediated perceptions. It increases the ability to discern among forms of environmentally encountered information. It values persistence and not just novelty. It stretches and extends the now, beyond the latest tweets, beyond the next business quarter, until the sense of the time period you inhabit exceeds the extent of your lifetime. It likewise expands the extent of where you think you live.

An ambient commons, if there ever were such a thing, would be quiet enough but seldom silent. In this alone it could change what it means to be here now. Like a quiet evening on the square, it might help recall how the noisy clamor of the morning's market was a means, and not an end in itself.

Notes

Prologue

1. Homage is due to Howard Rheingold's wonderful but different "Shibuya Epiphany," in *Smart Mobs: The Next Social Revolution* (New York: Basic Books, 2003), 1–28, and to Nelson Goodman's explanation of this otherwise seldom-used word, *replete*, in *Languages of Art: An Approach to a Theory of Symbols*, 2nd ed. (Indianapolis: Hackett, 1976), 229–235.

Chapter 1

1. News items about "glowing rectangles" proliferate after June 2009, when the satire magazine *Onion* published a fictional piece: "Report: 90% of Waking Hours Spent Staring at Glowing Rectangles." Whatever the actual numbers, they must be high. "Researchers were able to identify nearly 30 varieties of glowing rectangles that play some role throughout the course of each day. Among them: handheld rectangles, music-playing rectangles, mobile communication rectangles, personal work rectangles, and bright alarm cubes, which emit a high-pitched reminder that it's time to rise from one's bed and move toward the rectangles in one's kitchen." http://www.theonion.com/articles/report-90-of-waking-hours-spent-staring-at-glowing,2747/

2. Mark Prendergast, *The Ambient Century: From Mahler to Trance: The Evolution of Sound in the Electronic Age*, with a preface by Brian Eno (London: Bloomsbury Press, 2001).

3. Claude Debussy, as quoted in Prendergast, *The Ambient Century*, 10.

4. Ambient Planet marketing agency website, www.ambient-planet.com, as of February 2010.

5. Lisa Reichelt, "Ambient Intimacy," 2007, http://www.disambiguity.com/ambient-intimacy/. "Knowing these details creates intimacy," Reichelt explains. "(It also saves a lot of time when you finally do get to catch up with these people in real life!) It's not so much about meaning, it's just about being in touch."

6. Clive Thompson, "I'm So Totally, Digitally Close to You," *New York Times Magazine,* September 21, 2008. "Social scientists have a name for this sort of incessant online contact," Thompson wrote. "They call it 'ambient awareness.' It is, they say, very much like being physically near someone and picking up on his mood through the little things he does—body language, sighs, stray comments—out of the corner of your eye."

7. See Hiroshi Ishii and Brygg Ullmer, "Tangible Bits: Towards Seamless Interfaces," in *ACM CHI 1997*, 234–241.

8. Paul Dourish, *Where the Action Is: The Foundations of Embodied Interaction* (Cambridge, MA: MIT Press, 2001), 115. "Embodiment is the property of our engagement with the world that allows us to make it meaningful" (126). As well as anyone, Dourish's work has shown the way from phenomenology to ambient interface to urban informatics.

9. Dourish, Suchman, Weiser, Brown, and many others influential to embodied information were at Xerox PARC in the mid-1990s, when I got to be a (helpless but observant) visitor on sabbatical there.

10. See Dan Hill, "New Soft City," City of Sound, 2010, http://www.cityofsound.com.

11. See Gernot Böhme, "Atmosphere as the Fundamental Concept of a New Aesthetics," *Thesis Eleven* 36 (August 1993): 113–126.

12. Thanks to my colleague Steve Jackson for "circumstantial."

13. See Leo Spitzer, "Milieu and Ambiance: An Essay in Historical Semantics," *Philosophy and Phenomenological Research* 3, no. 2 (December 1942): 169–218.

14. See Steven Berlin Johnson, *The Invention of Air: A Story of Science, Faith, Revolution, and the Birth of America* (New York: Riverhead Books, 2008).

Ever a bellwether of digital culture, Johnson saw the relevance of a history of Joseph Priestly and the discovery of oxygen.

15. Spitzer, "Milieu and Ambiance," 2.

16. See note 13. This Spitzer essay may have been an esoteric find, but perhaps something from before the cybernetic age was just what the inquiry needed.

17. Spitzer, "Milieu and Ambiance," 182, citing Pliny the Elder, *Naturalis Historia, Book XIV,* line 11.

18. *Whitaker's Words,* a Latin dictionary, was used for this translation. Although competing versions and translations of Pliny exist, that is no reason to put aside this vivid image of vines.

Chapter 2

1. Epistemology is not my field. I have read many overview and seminal texts and formed my own naïve viewpoints from them. That process is the source of many apparent generalizations in this text. Wherever aware of a particular scholar behind an idea, I have made a citation. When stating my own take on what appears to be common wisdom with countless scholars behind it, I have attempted to trace its origins wherever practical to do so.

2. The crosswalk example came from an experience in Savannah, Georgia, where I gave a talk at the 2008 IxDA interaction design conference. Little did I know that Dan Saffer, the host who had just introduced me, had a crosswalk on the cover of his excellent book *Designing for Interaction: Creating Innovative Applications and Devices*. (San Francisco: New Riders, 2006). Tick, tick.

3. Albert Borgmann, *Holding On to Reality: The Nature of Information at the Turn of the Millennium* (Chicago: University of Chicago Press, 1999), 1.

4. Peter Lyman and Hal R. Varian, "How Much Information 2003?" http://www2.sims.berkeley.edu/research/projects/how-much-info-2003/.

5. "The Data Deluge: Businesses, Governments and Society Are Only Starting to Tap Its Vast Potential," *Economist*, February 25, 2010.

6. Eric Schmidt, as quoted in M. G. Sigler, "Eric Schmidt: Every 2 Days We Create As Much Information As We Did Up to 2003," Tech Crunch, August 4, 2010, http://techcrunch.com/2010/08/04/schmidt-data/.

7. David Levy, "Information Overload," in Kenneth Einar Himma and Herman T. Tavani, eds., *The Handbook of Information and Computer Ethics* (New York: Wiley, 2008), 510–512.

8. "You Choose: If You Can Have Everything in 57 Varieties, Making Decisions Becomes Hard Work: The Tyranny of Choice," *Economist*, December 16, 2010.

9. On digital natives, see Don Tapscott, *Growing Up Digital: The Rise of the Net Generation* (Cambridge, MA: Harvard Business School Press, 2000).

10. Consider the two poles of mindfulness and its disappearance. Winifred Gallagher, in *Rapt: Attention and the Focused Life* (New York: Penguin Press, 2009), has explained how rapt attention not only remains possible but can also be cultivated, both through reflection and through careful cultural production, and that in many cases neuroscientists can show this. At the other pole, Maggie Jackson, in *Distracted: The Erosion of Attention and the Coming Dark Age* (Amherst, NY: Prometheus Books, 2008), has confronted the myth of multitasking and the abdication of agendas to browsing, grazing, and feedburning, and, though Jackson is far from hopeful, she likewise sees possible practices to counter this. Of course, there are other approaches to unfettered attention, such as traditional Zen Buddhism.

11. Jean-Baptiste Michel, Yuan Kui Shen, Aviva Presser Aiden, Adrian Veres, Matthew K. Gray, William Brockman, the Google Books Team, Joseph P. Pickett, Dale Hoiberg, Dan Clancy, Peter Norvig, Jon Orwant, Steven Pinker, Martin A. Nowak, and Erez Lieberman Aiden, "Quantitative Analysis of Culture Using Millions of Digitized Books," Sciencexpress, December 16, 2010, http://www.librarian.net/wp-content/uploads/science-googlelabs.pdf.

12. Richard Saul Wurman, *Information Anxiety* (New York: Doubleday, 1989), 38, 211.

13. James's original aphorism reads: "My experience is what I agree to attend to." Another reads: "Every one knows what attention is." William James, *Principles of Psychology* (Cambridge, MA: Harvard University Press, 1890), 1:402, 403. See chapter 3 of this volume for more detailed excerpts and citations.

14. Herbert Simon, "Designing Organizations for an Information-Rich World," in *Computers, Communication, and the Public Interest*, ed. Martin Greenberger (Baltimore: Johns Hopkins University Press, 1971), 40–41.

15. Erasmus, as quoted in Ann Blair, "Information Overload, the Early Years" *Boston Globe*, November 28, 2010.

16. See Ann Blair, "Coping with Information Overload in Early Modern Europe," *Journal of the History of Ideas* 64 (2003): 11–28.

17. Blair, "Information Overload, the Early Years."

18. The argument that anyone in any era may have felt overload has to be balanced against the argument that artificial stimuli for which cognitive mechanisms have not evolved are ever more ubiquitous, cognitively engineered in their own ways, and habit-forming. See David Levy, "Information Overload," for a good place to start in on this debate.

19. See course description for "History of Information," C103, School of Information, University of California Berkeley, http://www.ischool.berkeley.edu/courses/103.

20. David Henkin, *City Reading: Written Words and Public Spaces in Antebellum New York* (New York: Columbia University Press, 1998), 101–136.

21. David Shenk, *Data Smog: Surviving the Information Glut* (San Francisco: Harper Collins, 1997), 29.

22. See video discussion between Lev Grossman (*Time*) and Reihan Salam (Daily Beast), "Information Obesity," Bloggingheads, August 6, 2010, http://video.nytimes.com/video/2010/08/06/opinion/1247468584580/bloggingheads-information-obesity.html.

23. See Geoff Nunberg, "Farewell to the Information Age," in *The Future of the Book* (Berkeley: University of California Press, 1997).

24. See Michael Buckland, "Information as Thing," *Journal of the American Society of Information Science* 42 (1991): 351–360.

25. Buckland, "Information as Thing," 359.

26. Luciano Floridi, "Semantic Conceptions of Information," *Stanford Encyclopedia of Philosophy*, ed. Edward N. Zalta (Winter 2007 Edition), http://plato.stanford.edu/archives/win2007/entries/information-semantic.

27. Floridi, "Semantic Conceptions of Information." Though never false in its original, natural form and location, nonsemantic information can, of course, be faked through human agency.

28. See Paul Grice, "Meaning." *Philosophical Review* 66 (1957): 377–388. A standard.

29. Luciano Floridi, Philosophy of Information, http://www.philosophyofinformation.net/blog/.

30. This section's many generalizations stem from having grown up on semiotics as a graduate student and young faculty member in schools of architecture.

31. For an exemplary work on the relationship of documentation, city form, and patterns of use, see Priscilla Ferguson, *Paris as Revolution: Writing the 19th Century City* (Berkeley: University of California Press, 1994).

32. Anyone who took part in the academy in the 1990s might find it difficult to sort out the literature on intrinsic information and sign theory from his or her own historical perspective and to state with any certainty which works were seminal. I, for one, read Roland Barthes in the French as an undergraduate. And, for quite some time, architecture students were all made to read Jacques Derrida.

33. Borgmann, *Holding On to Reality*, 218. Borgmann opens with similar examples of natural meaning: gravel beds indicate a river, cottonwoods its bank, and an osprey nest the presence of trout in the river (1). Borgmann's explanations of environmental knowing, especially through the contingency of built form, parallel my own thoughts and have given better voice to them. The whole point is that many of us share these same thoughts.

34. José Luis Borges, "Of Exactitude in Science," in *A Universal History of Infamy* (London: Penguin, 1975). The map larger than the territory long ago became a very findable meme online.

35. Borgmann, *Holding On to Reality*, 181, also uses this common example of music. But when you tell others that what they are listening to is a recording, you get much rolling of the eyes. For them, the file *is* the music.

36. Borgmann, *Holding On to Reality*, 221.

37. Borgmann, *Holding On to Reality*, 209.

38. Borgmann, *Holding On to Reality*, 22–23.

39. Borgmann, *Holding On to Reality*, 1.

40. Floridi, in "Semantic Conceptions of Information," calls "information 'for' reality" "instructional information."

41. Borgmann, *Holding On to Reality*, 113. "Contingency is the one concession thoughtful theorists like to make to the eloquence of the world" (106).

42. Jackson, *Distracted*, 25–26.

43. Bill McKibben, *The Age of Missing Information* (New York: Random House, 1992).

44. Borgmann, *Holding On to Reality*, 14.

45. Given what cognitive neuroscience has learned about the importance of knowledge representations and intentional states, a new round of "environment-and-intent" studies could do much to compensate for a previous, mid-twentieth-century round of overly deterministic "environment-and-behavior" studies.

46."Oh . . . , cool . . . , where?": I actually got this response from someone in my own place of work.

Chapter 3

1. Attention science is not my field. This inquiry is for others who are likewise curious. I have necessarily made generalizations from the literature, and do not claim to have introduced any ideas myself. My contribution is their juxtapositions with other fields here (see note 1 to chapter 2).

2. James, *Principles of Psychology*, 1:402.

3. See Bill McKibben, foreword to Jackson, *Distracted*, 9–10. Cultural critic McKibben put it well: "Distraction has always been a human condition. Sages have always been quick to point out how even a few minutes of meditation prove the jumpy nature of our consciousness—our monkey minds. But now every force conspires to magnify that inattentiveness: technology has made distraction ubiquitous" (9).

4. "Anything humans do," even sex? See, for example, http://scitech.blogs.cnn.com/2010/05/06/texting-during-sex-some-say-its-ok/.

5. See Eyal Ophir, Clifford Nass, and Anthony D. Wagner, "Cognitive Control in Media Multitaskers," *Proceedings of the National Academy of Sciences* 106 (2009): 15583–15587.

6. See Sherry Turkle, *Alone Together: Why We Expect More from Technology and Less from Each Other* (New York: Basic Books, 2011).

7. For Linda Stone on continuous partial attention, see http://lindastone.net/. An abundance of attention articles has been collected here, including even one on the possibility of "email apnea."

8. Linda Stone, "Continuous Partial Attention—Not the Same As Multi-Tasking," Bloomberg Businessweek Online, July 24, 2008, http://www.businessweek.com/business_at_work/time_management/.

9. Linda Stone, "Why Email Can Be Habit-Forming," Huffpost Healthy Living Blog, http://www.huffingtonpost.com/linda-stone/why-email-can-be-habit-fo_b_324781.html.

10. James, *Principles of Psychology,* 1:402–458.

11. James, *Principles of Psychology,* 1:402 (emphasis in original).

12. Jackson, *Distracted,* 24.

13. Jackson, *Distracted,* 72.

14. See, for example, Lawrence Barsalou, "Grounded Cognition," *Annual Review of Psychology* 59 (2008): 617–645.

15. Of the standard university textbooks on cognitive science, perhaps the most accessible overview is in John Anderson, *Cognitive Psychology and Its Implications,* 6th ed. (New York: Worth, 2004).

16. Posner's three attentional networks were first introduced over forty years ago. See Michael Posner and Stephen Boises, "Components of Attention," *Psychological Review* 78 (1971): 391–408.

17. Jackson, *Distracted,* 25, 239, 243, 247.

18. See Michael Posner, "Socializing the Young As Viewed from Natural and Social Science Perspectives," keynote address to the University of Michigan Center for Culture, Mind, and the Brain Annual Conference, April 2010.

19. See Michael Posner and Mary Rothbart, *Educating the Human Brain* (Washington, DC: American Psychological Association, 2006), for perhaps the best retrospect on Posner and Rothbart's work, and for prospects in training for mindfulness.

20. See Christopher Chabris and Daniel Simons, *The Invisible Gorilla: And Other Ways Our Intuitions Deceive Us* (New York: Crown, 2010). The experiment that forms the basis of this book was originally reported in Steven B.

Most, Daniel J. Simons, Brian J. Scholl, Rachel Jimenez, Erin Clifford, and Christopher F. Chabris, "How Not to Be Seen: The Contribution of Similarity and Selective Ignoring to Sustained Inattentional Blindness," *Psychological Science* 12 (2001): 9–17.

21. Jackson, *Distracted*, 237. Jackson calls Michael Posner "unarguably the greatest attention scientist of our time" (237).

22. Howard Pashler explained "attention set" especially well in *The Psychology of Attention* (Cambridge, MA: MIT Press, 1998), 167–216.

23. Of the standard university textbooks on cognitive science, perhaps the most accessible overview on vision is Douglas Medin, Brian H. Ross, and Arthur B. Markman, *Cognitive Psychology*, 3rd ed. (Fort Worth: Harcourt, 2001).

24. Anne Treisman and Garry Gelade, "A Feature-Integration Theory of Attention," *Cognitive Psychology* 12 (1980): 97–136.

25. On control of the hands, see one of my earlier books, *Abstracting Craft: The Practiced Digital Hand* (Cambridge, MA: MIT Press, 1996), 1–12.

26. See Joshua S. Rubinstein, David E. Meyer, and Jeffrey E. Evans, "Executive Control of Cognitive Processes in Task Switching," *Journal of Experimental Psychology, Human Perception, and Performance* 27 (2001): 763–797.

27. For an extensive nonspecialist account of the neuroscience of attention and of the widely cited neuroscientist of attention David Meyer in particular, see Jackson, *Distracted*. For a classic technical account, see Rubinstein, Meyer, and Evans, "Executive Control of Cognitive Processes in Task Switching."

28. Ophir, Nass, and Wagner, "Cognitive Control in Media Multitaskers"; see also Avi Solomon, "Eyal Ophir on the Science of Multitasking," BoingBoing, November 7, 2011, http://boingboing.net/.

29. Steven Pinker, "Mind over Mass Media," *New York Times*, June 10, 2010.

30. Pinker, "Mind over Mass Media," "Experience does not revamp the basic information-processing capacities of the brain" Pinker went on to write. "Speed-reading programs have long claimed to do just that, but the verdict was rendered by Woody Allen after he read *War and Peace* in one sitting: 'It was about Russia.'"

31. See Wendy Wood and Leona Tam, "Changing Circumstances, Disrupting Habits," *Journal of Personality and Social Psychology* 88 (2005): 918–933.

32. See Edward Hall, *The Hidden Dimension* (New York: Doubleday, 1966). In the 1960s, anthropologist Hall popularized ideas of "social distance," in which tacit knowledge exists as a game of inches, which he called "proxemics." Deskilling in those interpersonal distances is evident today, when social distances are more often mediated online and face-to-face encounters are becoming more awkward; for example, a retail clerk may now first speak to you at some strangely arbitrary range before making eye contact.

33. See Michael Polanyi, *Personal Knowledge: Toward a Post-Critical Philosophy* (Chicago: University of Chicago Press, 1958).

34. James Gleick, *Genius: The Life and Science of Richard Feynman* (New York: Vintage Books, 1993), 409; Andy Clark, *Supersizing the Mind: Embodiment, Action, and Cognitive Extension* (New York: Oxford University Press, 2008), xxv.

35. Stafford, *Echo Objects: The Cognitive Work of Images* (Chicago: University of Chicago Press, 2007), 10–14.

36. Stafford, *Echo Objects,* 1. The opening of Stafford's book deals with transdisciplinary challenges in the "cognitive revolution."

37. James, *Principles of Psychology*, 1:114.

Chapter 4

1. Although I am not a cognitive psychologist, and have inevitably made some generalizations from the literature (see note 1 to chapter 2), embodied cognition has been a thread through many of my writings before this one.

2. Andy Clark and David Chalmers, "The Extended Mind," *Analysis* 58 (1998):7–19. Reprinted in *Philosopher's Annual* 21 (1998): 59–74.

3. Andy Clark, *Being There: Putting Brain, Body and World Together Again* (Cambridge, MA: MIT Press, 1997), 83.

4. "You aren't your brain" has been said by any number of people, but, for a fine recent essay, see Alva Noë, "Art and the Limits of Neuroscience," *New York Times*, December 4, 2011.

5. Victor Kapetelinin and Bonnie Nardi, *Acting with Technology: Activity Theory and Interaction Design* (Cambridge, MA: MIT Press, 2006), 29–35.

6. Kapetelinin and Nardi, *Acting with Technology,* 69–70.

7. Kapetelinin and Nardi, *Acting with Technology,* 42.

8. Bonnie Nardi, "Activity Theory and Human-Computer Interaction," in *Context and Consciousness*, ed. Bonnie Nardi (Cambridge, MA: MIT Press, 1996), 14. My own earlier books looked at the nature of tools and the importance of intent as opposed to behavior. *Digital Ground: Architecture, Pervasive Computing, and Environmental Knowing* (Cambridge, MA: MIT Press, 2004) drew as much on the work of Nardi as on that of any other anthropologist. And, growing out of my first encounters by way of the cognition research community around Harvard with the work of the seminal Russian activity theorists Lev Vygotsky and Aleksei Leontiev, my 1996 book, *Abstracting Craft*, remains a good way into the kinds of attention that flow from skillful tool use.

9. Kapetelinin and Nardi, *Acting with Technology*, 196, closes on "postcognitivist" theories: activity theory, distributed cognition, actor-network theory, phenomenology, as a range of approaches that are "highly critical of mind-body dualism."

10. William Warren, "Perceiving Affordances: Visual Guidance of Stair Climbing," *Journal of Experimental Psychology: Human Perception and Performance* 10 (1984): 683–703.

11. For excellent explanations of habituation, automatic actions, and procedural chunks, see John Anderson's classic textbook, *Cognitive Psychology and Its Implications,* 6th ed. (New York: Worth, 2004).

12. I am grateful to my colleague Michael Cohen for his clear and penetrating study of routine, and for his many suggestions about other work cited here.

13. William James, *Principles of Psychology,* 1:121. "Habit a second nature! Habit is ten times nature, the Duke of Wellington is said to have exclaimed" (1:120).

14. A Google search for "runaway ontology" returned fewer than 10 items in January 2012.

15. Dourish, *Where the Action Is*, 21.

16. Antonio R. Damasio, *Descartes' Error: Emotion, Reason, and the Human Brain* (New York: Putnam's, 1994).

17. Melvin Goodale and David Milne, "Separate Visual Pathways for Perception and Action," *Trends in Neuroscience* 15 (1992): 20–25.

18. I am grateful to many of my colleagues at the University of Michigan, itself a situated knowledge community of the highest order, for sharing their expertise on the situatedness of expertise and the potential cost of excess mediation.

19. "Situational awareness" has been a part of the "sensemaking" conversations in my university (see note 11 to chapter 5). I cannot say where I first was around people making use of this expression.

20. Clark, *Being There*, 46.

21. See Lev Vygotsky, *Thought and Language*, rev. ed. (Cambridge, MA: MIT Press, 1986). Andy Clark gave ample interpretation of Vygotsky's thinking in *Being There*. I first came across Vygotsky in the early 1990s and continue to find his version of activity theory extremely useful.

22. My 1996 book, *Abstracting Craft*, explored the relations of tool, medium, practice, and intent.

23. Clark, *Being There*, 45.

24. Clark, *Being There*, 83.

25. My 2004 book, *Digital Ground*, explored intentionality in habitual contexts.

26. Böhme, "Atmosphere as the Fundamental Concept of a New Aesthetics," 125.

27. See Mihaly Csikszentmihalyi, *Flow: The Psychology of Optimal Experience* (New York: Harper and Row, 1990).

28. Rachel and Stephen Kaplan are generally credited for the now widely used expression "effortless attention." See especially their book *The Experience of Nature: A Psychological Perspective* (New York: Cambridge University Press, 1989).

29. Brian Bruya, *Effortless Attention: A New Perspective in the Cognitive Science of Attention and Action* (Cambridge, MA: MIT Press, 2010): 3–29.

30. Bruya, *Effortless Attention*, 11.

31. Bruya, *Effortless Attention*, 12.

32. Stephen Kaplan, "The Restorative Benefits of Nature: Toward an Integrative Framework," *Journal of Environmental Psychology* 15 (1995): 169–182.

33. See Rita Berto, "Exposure to Restorative Environments Helps Restore Attentional Capacity," *Journal of Environmental Psychology* 25 (2005): 249–259.

34. See Peter Kahn et al., "A Plasma Display Window? The Shifting Baseline Problem in a Technologically Mediated Natural World," *Journal of Environmental Psychology* 28 (2008): 192–199.

35. Bruya, *Effortless Attention*, 5.

36. Gallagher, *Rapt*, 2.

37. Gallagher, *Rapt*, 9. "Along with the Apollonian task of organizing your world," Gallagher went on to write, "attention enables you to have the kind of Dionysian experience beautifully described by the old-fashioned term *rapt*—completely absorbed, engrossed, fascinated, perhaps even "carried away" that underlies life's deepest pleasures, from the scholar's study to the carpenter's craft to the lover's obsession" (9–10).

Chapter 5

1. William J. Mitchell, "In the First Place," in *Placing Words: Symbols, Space, and the City* (Cambridge, MA: MIT Press, 2005), 8. In my long-term work of picking up after the great master, this particular piece was an inspiration, more so than his more influential work on smart cities.

2. Mitchell, "In the First Place," 3.

3. Mitchell, "In the First Place," 4.

4. Mitchell, "In the First Place," 9.

5. Mark Twain, "A True Story, Repeated Word for Word As I Heard It," *Atlantic Monthly*, November 1874, 591–594. Reprinted in the sesquicentennial anthology Robert Vare and Daniel B. Smith, eds., *The American Idea: The Best of the* Atlantic Monthly (New York: Atlantic Monthly, 2007), 28–33.

6. Mitchell, *Placing Words*, 13.

7. My 2004 book, *Digital Ground*, likewise addressed habit, but only with respect to reestablishing embodiment and pattern for digital design and not

with the present emphasis on the economics of attention. A few remarks from "Habitual Contexts," chapter 3 of that book, are echoed here.

8. As cited in *Digital Ground*, interactivity scholar Paul Dourish has observed: "At the heart of tangible computing is the relationship between activities and the space in which they are carried out. Tangible computing expands this in three ways: through the configuration of space, through the relationship of body to task, and through physical constraints." Dourish, *Where the Action Is*, 158.

9. See Victor Kapetelinin and Bonni Nardi, *Acting with Technology: Activity Theory and Interaction Design* (Cambridge, MA: MIT Press, 2006). For detailed citations, see chapter 4 of this volume.

10. Thus at Xerox PARC in the 1990s, where the ideas of ubiquitous computing first took form, there was active work by Daniel Russell, Peter Pirolli, and colleagues on sensemaking and by Lucy Suchman and colleagues on situated actions. See, for example, Daniel M. Russell, Mark J. Stefik, Peter Pirolli, and Stuart K. Card, "The Cost Structure of Sensemaking," in *InterCHI 1993*, 269–276.

11. Karl Weick, "Sensemaking," in *International Encyclopedia of Organizational Studies* (Thousand Oaks, CA: Sage, 2008), 4:1403–1406. See also Karl E .Weick, Kathleen M. Sutcliffe, and David Obstfeld, "Organizing and the Process of Sensemaking," *Organization Science* 16 (2005): 409–421.

12. I am grateful to my colleague Erik Hofer for suggesting this everyday example.

Chapter 6

1. Claudia Walde, *Sticker City:Paper Graffiti Art* (New York: Thames and Hudson, 2007), 38.

2. David Pescovitz, "Graffiti at the National Portrait Gallery," February 20, 2008, http://boingboing.net/2008/02/20/graffiti-at-the-nati.html.

3. Dan Witz, as quoted in Walde, *Sticker City*, 51.

4. My own experience of New York began and was mainly shaped in the 1970s.

5. See Joe Austin, *Taking the Train: How Graffiti Art Became an Urban Crisis in New York City* (New York: Columbia University Press, 2001).

6. James Wilson and George Keiling, "Broken Windows," *Atlantic Monthly*, March 1982.

7. See iwishthiswas.cc.

8. Walde, *Sticker City*, 9.

9. Walde, *Sticker City*, 38.

10. As of January 2012, the augmented reality (AR) business appeared ready to explode. Google's fall 2001 announcement of plans for Google Glasses, a mass-market wearable AR overlay platform in the form of stylish eyeglasses, had much to do with this. Former Nokia director Tomi Ahonen proclaimed AR "the eighth mass medium," and predicted a billion users within a decade. (See http://tedxtalks.ted.com/video/TEDxMongKok-Tomi-Ahonen-Augment). Elsewhere, MetaIO, a leading European AR company, posted that, "According to Gartner Inc., Augmented Reality is one of the Top 10 strategic IT technologies of our time. And Juniper Research forecasts $1.5 billion revenue stream by 2015. (See http://www.metaio.com/company/company/). Under these circumstances, this slower-moving work of print, on its way to press, cannot say much, but can hope for increased (if indirect) relevance.

11. In information design, and with possible lessons for new forms of urban inscriptions, there follows a debate about bottom-up versus top-down forms. Classification schemes traditionally were top-down and sought to be comprehensive on their designated topics. A taxonomy (such as a library would use to shelve this book in exactly one place), is a top-down, tree-structured hierarchy. Or, for nonhierarchical classification, an information designer might develop an *ontology*, an existential term appropriated from metaphysics to deal with markup languages' capacity for bottom-up schemas.

12. Bradley H. McLean, *An Introduction to Greek Epigraphy of the Hellenistic and Roman Periods from Alexander the Great until Constantine* (Ann Arbor: University of Michigan Press, 2002), 2.

13. Victor Hugo, *The Hunchback of Notre Dame*, book V, chapter 2, "This Will Kill That," as translated by Isabel F. Hapgood in 1861.

14. See Henkin, *City Reading*, 69–136, on the many forms of print used at street level.

15. Taubman College at the University of Michigan, my own place of work, operates a robotic water jet for cutting slab stone.

16. Ferguson, *Paris as Revolution,* 16.

17. Ferguson, *Paris as Revolution,* 29.

18. Ferguson, *Paris as Revolution,* 39.

19. Society of Environmental Graphic Design, www.segd.org.

20. National Gallery of Scotland. Signspotting exhibition, summer 2009.

21. Erik Davis, "Urban Markup Language," *Wired*, September 2004.

22. Photography section headings, Eric Sadin, *Times of the Signs* (Berlin: Springer, 2003).

23. See http://www.graffitiresearchlab.com/blog/. A Google search for "LED throwies" yielded more than 100,000 hits in February 2009.

24. Ethan Todras-Whitehill, "Making Connections Here and Now," *New York Times*, January 25, 2006.

25. Camille Johnson, "Yellow Arrow Aimed at Building Art Community," *Harvard Crimson,* November 19, 2004.

26. Daily RFID press release, December 2008, http://www.rfid-in-china.com/products_667_1.html: "DAILY RFID has unveiled mini RFID Sticker Tag-08 (measured in dia. 9 mm) for use in most RFID applications, including directly on metallic surfaces. This adhesive RFID Sticker tag can easily be affixed to on flat and clean surfaces with its 3M Glue."

27. See Avery Dennison Inc.'s website: http://rfid.averydennison.com/

28. Bruce Sterling, "arphid watch," http://www.wired.com/beyond_the_beyond/category/arphid-watch/.

29. See the Situated Technologies pamphlet series, edited by Omar Kahn, Trevor Scholz, and Mark Shepard, published by the Architecture League of New York, and begun in 2006.

30. Richard Sennett, *The Conscience of the Eye: The Design and Social Life of Cities* (New York: Norton, 1990), 205.

Chapter 7

1. See www.displaybank.com. Displaybank is a leading market research consultancy in display and photovoltaic panels.

2. See Anna McCarthy, *Ambient Television: Visual Culture and Public Space* (Durham: Duke University Press, 2002), whose cover image has just that juxtaposition.

3. See Hannah Higgins, *The Grid Book* (Cambridge, MA: MIT Press, 2007).

4. David Nye, *Electrifying America: Social Meanings of a New Technology* (Cambridge, MA: MIT Press, 1992), 34–41.

5. Walter Benjamin, "One Way Street: This Space for Rent," in *Selected Writings*, vol. 1 (1913–26), ed. Marcus Bullock and Michael W. Jennings, trans. Rodney Livingstone (Cambridge, MA: Harvard University Press, 1996), 476.

6. References to Pittsburgh here and in later chapters come from my having grown up there.

7. Neal Stephenson, *Snow Crash* (New York: Bantam, 1992), 5.

8. See Christa Van Santen, *Light Zone City: Light Planning in the Urban Context* (Basel: Birkhauser, 2006).

9. The existence of a signage district implies restraint elsewhere in the city. Chaotic illuminations can produce distress among those who must live with them. To many, the experience of a huge, full-motion video facade is especially uncomfortable, suggesting that something that large needs more minimalist programming. Thus, as a revolution in display technology diversifies the scales and resolutions of surfaces that glow, it invites an environmental history and criticism.

10. Among signage aficionados, there exists no better indicator of America's overall cultural decline than the replacement (made infamous in Gary Hustwit's 2007 film *Helvetica*) of a classic space-age motel sign in Sedalia, Missouri, by a generic boxed backlit sign with tacky typography and gratuitous graduated background fill.

11. While on sabbatical in Berkeley and writing an earliest draft of *Ambient Commons* in January 2010, I could watch the billboard on the Bay Bridge approach change images, from several miles away in the hills.

12. KCET public television station, "Billboard Confidential," October 9, 2008, www.kcet.org/shows/socal_connected/content/environment/billboards/

13. Louis M. Brill, "Media Facades: High-Tech Building Wraps," SignIndustry.com, February 16, 2009, http://www.signindustry.com/outdoor/

articles/2009-02-16-LB-LED_Media_Facades_High-Tech_Digital_Building_Wraps.php3.

14. www.mediaarchitecture.org/kings-road-tower/.

15. Valenta Croci, "Dynamic Light: The Media Facades of realities:united," *Architectural Design* 80, no. 1 (January–February 2010): 136–139.

16. Carlos Ferré, "INSIGHT: Iconic Architecture in the Digital Age," ArchNewsNow.com, March 23, 2010, http://www.archnewsnow.com/features/Feature324.htm. Ferré's company A2aMedia introduced MediaMesh to North America.

17. Le Corbusier, *Vers une architecture* (Paris: G. Crès, 1924), "The history of architecture is the history of the struggle for light, the struggle for windows."

18. On facades as screens, see Mark C. Taylor, *The Moment of Complexity: Emerging Network Culture* (Chicago: University of Chicago Press, 2001), "From Grid to Network," 19–46. "Here it becomes undeniable that 'business art *is,*' as Warhol insists, 'the step that comes after Art,'" 46

19. Anne Friedberg, *The Virtual Window: From Alberti to Microsoft* (Cambridge, MA: MIT Press, 2006).

20. Leon Battista Alberti, as quoted in Friedberg, *The Virtual Window*, 27.

21. Friedberg, *The Virtual Window*, 101–149, 192.

22. Friedberg, *The Virtual Window*, 78.

23. Friedberg, *The Virtual Window*, 80.

24. On the composite as the most characteristic operation of digital media, see Lev Manovich, *Language of New Media* (Cambridge, MA: MIT Press, 2002), 136–161.

25. How odd for a technology whose advantage is the absence of backlit glow to make a brand of the act of lighting up: Kindle®.

26. "Magink enables an indoor video to be enhanced and blended into the surrounding environment without dominating, as other light emitting technologies do," http://catalogs.infocommiq.com/avcat/ctl4552/index.cfm?manufacturer=magink-display-technologies&product=magink-indoor.

27. Transflective™ is a trademark of Pixel Qi.

28. Susanne Seitinger, "Liberated Pixels: Alternative Narratives for Lighting Future Cities," Ph.D. diss., MIT Media Lab, 2010.

29. William J. Mitchell, "Smart City 2020: Emerging Technologies Are Poised to Reshape Our Urban Environments," *Metropolis,* March 20, 2006, http://www.metropolismag.com/story/20060320/smart-city-2020.

30. Giulio Jacucci, Ann Morrison, Gabriela Richard, Jari Kleimola, Peter Peltonen, Lorenza Parisi, and Toni Laitinen, "Worlds of Information: Designing for Engagement at a Public Multi-touch Display," in *ACM CHI 2010*, 2267–2276. See also Peter Peltonen, Esko Kurvinen, Antti Salovaara, Giulio Jacucci, Tommi Ilmonen, John Evans, Antti Oulasvirta, and Petri Saarikko, "It's Mine, Don't Touch!: Interactions at a Large Multi-touch Display in a City Center," in *ACM CHI 2008*, 1285–1294.

31. Peter Gall Krogh, Martin Ludvigsen, Andreas Lykke-Olesen, Aarhus School of Architecture and Kaspar Rosengreen Nielsen, Computer Science, University of Aarhus, iFloor, Aarhus Municipal Library. http://www.interactivespaces.net/news/news.php?newsId=26.

32. It is especially regrettable when project managers value-engineer ceilings down to cheaper and more generic materials. One memorable instance of this is the arrivals hall in Austin-Bergstrom Airport (opened in 1999), where the ornamental ceiling in the originally accepted design was value-engineered into a nondescript plain one.

33. My 2004 book, *Digital Ground*, explored these mental mappings of place.

34. Clay Shirky, *Here Comes Everybody: The Power of Organizing without Organizations* (New York: Penguin, 2009), 81.

35. See Anna Klingman, *Brandscapes: Architecture and the Experience Economy* (Cambridge, MA: MIT Press, 2007).

Chapter 8

1. Stephen Berlin Johnson, *The Invention of Air: A Story of Science, Faith, Revolution, and the Birth of America* (New York: Riverhead Books, 2008), 87–94. Priestley built a means to de-phlogistinate air. Only later did Lavoisier come up with the name still used for what Priestley discovered: oxygen.

2. MIT Project Oxygen. http://oxygen.csail.mit.edu/.

3. Interlink Research Consortium, "State of the Art in Ambient Computing and Communication Environments," Information Society Technology Advisory Group (ISTAG), advisory body to the European Commission, 2008, http://interlink.ics.forth.gr/ConsultationForum/Files/InterLink_WG2_State_of_the_Art_Report.pdf.

4. Bruno Latour, "Air," in Caroline A. Jones, ed., *Sensorium: Embodied Experience, Technology, and Contemporary Art* (Cambridge, MA: MIT Press, 2006), 104–107.

5. J. M. W. Turner retrospective, New York Metropolitan Museum of Art, 2008. http://www.metmuseum.org/exhibitions/jmw_turner/more.asp.

6. Sergio Leone, *Once Upon a Time in the West*, 1968. Sound by Ennio Morricone.

7. Böhme, "Atmosphere as the Fundamental Concept of a New Aesthetics," 116.

8. Böhme, "Atmosphere," 116.

9. "And aura is clearly something which flows forth spatially," Böhme wrote of Benjamin walking in the shadow of mountains, "almost something like a breath or a haze—precisely an atmosphere. Benjamin says that one 'breathes' the aura." Böhme, "Atmosphere," 117.

10. Prendergast, *The Ambient Century*, 8–11. "For many modern music began with the 'voluptuous ambience' of *Prélude à l'après-midi d'un faune* (*Prelude to the Afternoon of a Faun*), composed in (1894)" (8). Debussy's most famous aphorism involved air, incidentally: "The century of airplanes deserves a music of its own."

11. Robert Morris, "Notes on Sculpture, 1-3," *Artforum* (4:6), February 1966: 42–44; (5:2), October 1966: 20–23; (5:10) June 1967: 24–29.

12. Stafford, *Echo Objects*, 31.

13. My colleague Amy Kulper has been very helpful in explaining this generational identity in atmospheric design.

14. Cynthia Zarin, "Seeing Things: The Art of Olafur Eliasson," *New Yorker*, November 13, 2006. "This year, Eliasson has had twelve solo shows around the world. . . . Until recently, Eliasson said yes to most of the projects he was offered. Now he accepts just one percent."

15. Zarin, "Seeing Things."

16. Husserl establishes a core for the phenomenology that the human-computer interaction discipline has built toward tangible interfaces. See Dourish, *Where the Action Is.*

17. Zarin, "Seeing Things."

18. For my part, I lay there until closing time, about two hours in all.

19. Writing on the embodied experience in The Weather Project, Stafford observed: "One might have thought that mirror neurons—only visually activated when another agent is observed acting in purposeful ways with other agents or things—would clarify this existential ambiguity. But as Eliassson's installation brilliantly shows, it does not." Stafford, *Echo Objects,* 134.

20. Philippe Rahm, audio interview with *Domus* magazine, 2009. http://domusweb.it/en/architecture/philippe-rahm-audio-interview/.

21. Philippe Rahm, interview with Archinect blog, http://archinect.com/features/article/96362/philippe-rahm-part-1.

22. Peter Zumthor, *Atmospheres* (Basel: Birkhauser, 2006). Zumthor's nine sensibilities and approaches are as follows:

The body of architecture: "the material presence of things in a piece of architecture; its frame" (21) "Not the idea of the body—the body itself!" (23).

Material compatibility: "An extraordinary sense of the presence and weight of materials" (28).

The sound of a space: "Listen! Interiors are like large instruments, collecting sound, amplifying it, transmitting it elsewhere" (29).

The temperature of a space: "It is well known that materials more or less extract the warmth from our bodies" (33).

Surrounding objects: "I'm impressed by the things people keep around them" (35).

Between composure and seduction: "The feeling that I am not being directed but can stroll at will—just drifting along, you know?" (43).

Tension between interior and exterior: "An incredible sense of place, an unbelievable feeling of concentration when we suddenly become aware of being enclosed, of something enveloping us, keeping us together, holding us, whether we be many or single" (47).

Levels of intimacy: "It all has to do with proximity and distance" (49).

The light on things: "To plan the building as a pure mass of shadows then, afterwards, to put in light as if you were hollowing out the darkness, as if the light were a new mass seeping in. . . . To go about lighting materials and surfaces systematically and to look at the way they reflect the light" (59).

23. Zumthor, *Atmospheres,* 42–44.

24. Luis Fernández-Galiano, *Fire and Memory: On Architecture and Energy* (Cambridge, MA: MIT Press, 2000), 4–6, 66.

25. Fernández-Galiano, *Fire and Memory*, 213.

26. The expression "conditioned by air" has been in the air, with each of us making our own reexpressions of it, after Peter Sloterdijk. The expression "air-condition" popularized this, after Bruno Latour, "Air-Condition: Our New Political Fate," *Domus* (868), March 2004.

27. On a warming planet, demand for cooling technology might only increase, even as the electricity it consumes contributes to still more warming.

28. Marsha Ackerman, *Cool Comfort: America's Romance with Air-Conditioning* (Washington, DC: Smithsonian Institution Press, 2002), 35, 44. Several advertisements from the Carrier Corporation highlight this excellent social history.

29. Michelle Addington, "Good-Bye, Willis Carrier" (1997), reprinted in Kim Tanzer and Rafael Longoria, eds., *The Green Braid* (New York: Routledge, 2007), 160.

30. Le Corbusier, as quoted in Addington, "Good-bye, Willis Carrier," 160.

31. Reyner Banham, *The Architecture of the Well-tempered Environment.* (London: Architectural Press, 1969), 209–212.

32. Banham, *Well-tempered Environment*, 12, 11.

33. Banham, *Well-tempered Environment*, 187. This passage is highlighted in my vintage copy from undergraduate years, amid the first oil crisis of the 1970s.

34. Lisa Heschong, *Thermal Delight in Architecture* (Cambridge, MA: MIT Press, 1979), 17–19.

35. Addington, "Good-bye, Willis Carrier," 160.

36. Henry Miller, *The Air-Conditioned Nightmare*, 1945 (San Francisco: New Directions, 1970).

37. Peter Jon Lindberg, "Bad Music in Public Spaces: With More and More Hotels, Restaurants, and Retailers Adopting Music as a Branding Device, *T+L* Sounds Off on How Their Choices Speak Volumes," *Travel and Leisure*, November 2009. Byline: "Peter Jon Lindberg, Travel + Leisure's editor-at-large, never leaves home without noise-canceling headphones."

38. Lindberg, "Bad Music in Public Spaces." On stress, Lindberg cited the Argentine-born conductor Daniel Barenboim, who found "the creep of background music into every corner of public life, calling it 'as disturbing [as] the most despicable aspect of pornography.'"

39. Joseph Lanza, *Elevator Music: A Surreal History of Muzak®, Easy-Listening, and Other Moodsong®* (Ann Arbor: University of Michigan Press, 2004), 207.

40. Lanza, *Elevator Music,* 39–40.

41. Lanza, *Elevator Music,* 22, 28.

42. Lanza, *Elevator Music,* 206. More trippy still: a string version of Iron Butterfly's "In-Na-Gadda-Da-Vida" (1969) also exists.

43. Lanza, *Elevator Music,* 3.

44. Drone Zone, founded by Rusty Hodge, was the first station (streaming since February 2000) on San Francisco–based, listener supported somafm.com: "Served best chilled, safe with most medications. Atmospheric textures with minimal beats."

45. Lindberg, "Bad Music in Public Spaces."

46. Lanza, *Elevator Music*, 212.

47. Ove Arup Associates, "San Francisco Federal Building," 2006, http://www.arup.com/Projects/San_Francisco_Federal_Building.aspx.

48. Randy Shaw, "San Francisco's Green Building Nightmare," 2008, www.beyondchron.org/news/ ("The verdict on the structure's function as an office space for federal employees is nearly unanimous: it is a disaster"); Ove Arup Associates, "San Francisco Federal Building," 2006 ("By forgoing a mechanical cooling system, the GSA was able to save U.S.$11m in construction costs and cites annual operational savings of U.S.$500,000").

49. Norbert A. Streitz, "Ambient Computing and the Disappearing Computer," in Norbert A. Streitz, Jörg Geissler, and Torsten Holmer, eds., *Ambiente: Workspaces of the Future*, 1998, 1. See also Norbert A. Streitz, Jörg

Geissler, and Torsten Holmer, "Roomware for Cooperative Buildings: Integrated Design of Architectural Spaces and Information Spaces," in Norbert A. Streitz, Shin'ichi Konomi, and Heinz-Jürgen Burkhardt, eds., *Cooperative Buildings: Integrating Information, Organization, and Architecture*, Proceedings of the First International Workshop (CoBuild '98) (Berlin: Springer, 1998), 4–21.

50. "IBM Smarter Buildings Survey White Paper," 2010, www-03.ibm.com/press/.

51. Thanks to Bruce Nordman of Lawrence Berkeley Laboratory for insightful conversation on the costs and benefits of designing interfaces for increased participation in indoor comfort.

52. Raymond Cole and Zosia Brown, "Reconciling Human and Automated Intelligence in the Provision of Occupant Comfort," *Intelligent Buildings International* 1 (2009): 39–55.

53. Hill, "New Soft City."

Chapter 9

1. Adam Greenfield, "The City Is Here for You to Use," 2009, unpublished manuscript

2. The United Nations defines a megacity as any urban agglomeration with over 10 million inhabitants. Richard Saul Wurman's 19.20.21 project, named for nineteen cities of 20 million inhabitants in the twenty-first century, aims to introduce comparative standards for urban data mapping. See www.192021.org.

3. A Lexis-Nexis word search for "smart cities" in world news media over the last thirty years yielded approximately 1,000 results in February 2011, two-thirds from the last ten years.

4. A Lexis-Nexis word search for "smart grid" in world news media over the last thirty years yielded approximately 1,000 results in February 2011; half from the last four years, and the earliest from 2003.

5. A word search for "urban computing" on the ACM digital library, the standard among databases of information technology research yielded only about 150 results in December 2010, almost all since the 2006 *IEEE Pervasive Computing* conference, with the first repeat use apparently by Eric Paulos.

6. Even by February 2011, there were only a handful of news search results for "urban informatics" on Lexis-Nexis; in February 2011 there were only about 50 news search results on the ACM digital library, none before 2007, with the first repeat use apparently by Marcus Foth.

7. See Howard Rheingold, "Cities, Swarms, Cell Phones: The Birth of Urban Informatics," September 2003, http://www.thefeaturearchives.com/topic/Culture/Cities__Swarms__Cell_Phones__The_Birth_of_Urban_Informatics.html. Reviewing work by Anthony Townsend and Steve Johnson.

8. See William J. Mitchell, *City of Bits: Space, Place, and the Infobahn* (Cambridge, MA, MIT Press, 1995); *e-topia: "Urban Life, Jim—but Not As We Know It"* (Cambridge, MA: MIT Press, 1999); and *Me++: The Cyborg Self and the Networked City* (Cambridge, MA: MIT Press, 2003).

9. William J. Mitchell, "Smart City 2020: Emerging technologies are poised to reshape our urban environments," *Metropolis*. March 20, 2006. "As the cities and their components become smarter, they begin to take new shapes and patterns. They become programmable. And the design of their software becomes as crucial—socially, economically, and culturally—as that of their hardware."

10. See SENSEable City Laboratory, MIT, main web page, April 2011, senseable.mit.edu.

11. Dan Hill, "Towards a New Architect: An Interview with Carlo Ratti," City of Sound, March 9, 2009, http://www.cityofsound.com/blog/2009/07/towards-a-new-architect-an-interview-with-carlo-ratti.html.

12. Mike Kuniavsky, "Information Is a Material," in *Smart Things: Ubiquitous Computing User Experience Design* (San Francisco: Morgan Kaufmann, 2010), 44. "Just as most consumers do not spend much time differentiating which parts of a device are made of glass, metal or silicon, they do not spend much time identifying which effects they experience are created by hardware, which by software and which by services. They see the device as a single thing" (46).

13. Interlink Research Consortium, "State of the Art in Ambient Computing and Communication Environments," Information Society Technology Advisory Group (ISTAG), advisory body to the European Commission, 2008, http://interlink.ics.forth.gr/ConsultationForum/Files/InterLink_WG2_State_of_the_Art_Report.pdf. This initiative foresaw a move toward a

"substrate" of information, similar to the usual "Internet of things" idea. It follows from an earlier EU–funded proactive initiative, the Disappearing Computer.

14. Susanne Dirks, Constantin Gurdgiev, and Mary Keeling, "Smarter Cities for Smarter Growth: How Cities Can Optimize Their Systems for the Talent-Based Economy," IBM Institute for Business Value, Executive Report 2010, June 29, 2010, 15, http://www.zurich.ibm.com/pdf/isl/infoportal/IBV_SC3_report_GBE03348USEN.pdf.

15. See http://www.aqueousadvisors.com/blog/?p=641.

16. See "Street Computing: Workshop Proceedings," Melbourne, November 2009, http://eprints.qut.edu.au/31086/1/31086.pdf.

17. "Street Computing: Workshop Proceedings."

18. Marcus Foth, Eric Paulos, Christine Satchell, and Paul Dourish, "Pervasive Computing and Environmental Sustainability: Two Conference Workshops," *IEEE Pervasive Computing* 8 (January 2009): 80, http://eprints.qut.edu.au/17164/1 /c17164.pdf.

19. Evgeny Morozov, "Death of the Cyberflaneur," *New York Times*, February 4, 2012.

20. See Simon Sadler, *The Situationist City* (Cambridge, MA: MIT Press, 1998).

21. Guy Debord, *On the Passage of a Few Persons through a Rather Brief Unity of Time*, 1959, film.

22. For what is perhaps still the most accessible work on interpersonal distance, see Hall, *The Hidden Dimension.*

23. Hill, "Towards a New Architect."

24. Richard Sennett, *The Craftsman* (New Haven: Yale Press, 2008), 286–296.

25. See lirneasia.net. "Bottom of the pyramid" offends as an expression, for it implies dominance by a small few elsewhere, uniformity of goals and processes, and benefit to the pyramid itself as the main motive for development.

26. See www.quividi.com.

27. One such target audio beam technology is Audio SpotLight™. See http://www.holosonics.com.

28. As of March 2011, outside.in had been acquired by AOL, to complement its (more suburban) hyperlocal news service, Patch. See http://www.stevenberlinjohnson.com/2011/03/aol-oi.html

29. Marcus Foth, Laura Forlano, Christine Satchell, and Martin Gibbs, eds., *From Social Butterfly to Engaged Citizen: Urban Informatics, Social Media, Ubiquitous Computing, and Mobile Technology to Support Citizen Engagement* (Cambridge, MA: MIT Press, 2012).

30. Beatrice da Costa et al., Pigeon Blog, 2006. See http://www.beatrizdacosta.net/pigeonblog.php.

31. See www.noisetube.net.

32. Eric Paulos, R. J. Honicky, and Ben Hooker, "Citizen Science: Enabling Participatory Urbanism," in Marcus Foth, ed., *Handbook of Research on Urban Informatics* (Hershey, PA: Information Science Reference, 2009), 419.

33. Paulos, Honicky, and Hooker, "Citizen Science."

34. Jong-Sung Hwang, "u-City: The Next Paradigm of Urban Development," in Foth, ed., *Handbook of Research on Urban Informatics*, 371.

35. "For Cities to Become Truly Smart, Everything Must Be Connected," *Economist,* special issue, "Living on a Platform," November 4, 2010.

36. Sara Corbett, "Can the Cellphone Help End Global Poverty?" Profile of Jan Chipchase, *New York Times Magazine,* April 13, 2008, http://www.nytimes.com/2008/04/13/magazine/13anthropology-t.html?_r=1.

37. Paul Dourish and Genevieve Bell, "The Infrastructure of Experience and the Experience of Infrastructure: Meaning and Structure in Everyday Encounters with Space," *Environment and Planning B* 34 (2007): 418. "Infrastructure is analytically useful, both because it is embedded into social structures, and because it serves as a structuring mechanism in itself" (418). Here is an argument for intrinsic information.

38. Elinor Ostrom, *Governing the Commons: The Evolution of Institutions for Collective Action* (Cambridge: Cambridge University Press, 1990), 214. In contrast, the top-down models of a less networked era too often assume "complete information, independent action, perfect symmetry of interests, no human errors, no norms of reciprocity, zero monitoring enforcement costs" (191).

39. Julian Bleecker and Nicolas Nova, "A Synchronicity: Design Fictions for Asynchronous Urban Computing," Situated Technologies, 2009 (Architecture League of New York Pamphlets, 5), http://www.situatedtechnologies.net/files/ST5-A_synchronicity.pdf.

40. Charlie Michaels and Bird (sign painter), the Night Sky Billboard Project, 2011, http://vimeo.com/22428957. In Detroit, you should still be able to see maps of the project's start displayed on unrented billboards.

41. See David Bollier, "Growth of the Commons Paradigm," in Charlotte Hess and Elinor Ostrom, eds., *Understanding Knowledge as a Commons: From Theory to Practice* (Cambridge, MA: MIT Press, 2007), 27–40.

42. Georg Simmel, "The Metropolis and Mental Life" (1903), in *The Sociology of Georg Simmel,* ed. and trans. Kurt Wolf (New York: Free Press, 1950), 4 (413).

43. Jonathan Crary, *Suspensions of Perception: Attention, Spectacle, and Modem Culture,* (Cambridge: MIT Press, 1999), 49.

44. Crary, *Suspensions of Perception,* 13.

45. Georg Simmel, "The Metropolis and Mental Life,"1.

46. Simmel, "The Metropolis and Mental Life." This from the translation most often found online, as at the Blackwell site that comes up first in search (in January 2010); but it is not clearly attributed. Is it Levine 1976? Others have used the earlier Wolf translation by which the essay was first brought back into regular use in the United States: "the intensification of nervous stimulation, resulting from the rapid crowding of changing images, the sharp discontinuity in the grasp of a single glance, and the unexpectedness of onrushing impressions. These are the psychological conditions which the metropolis creates." For instance, Michael Hays used that one in an essay that helped revive interest. "Critical Architecture: Between Culture and Form," *Perspecta* 21 (1984).

47. Simmel, "The Metropolis and Mental Life," 4.

48. See David Frisby, *Georg Simmel* (London: Tavistock, 1984). Early in the work of postmodern social criticism, Frisby revived Simmel, who unlike his contemporaries Durkheim and Weber had been largely forgotten.

49. "The focus on everyday life, the essayistic style, and the general resistance to master schemas that made Simmel unattractive to high modern sociologists made him ideal to the postmodernists." Frisby, *Georg Simmel.*

50. "Extremities and peculiarities and individualizations must be produced and they must be over exaggerated merely to be brought into the awareness of even the individual himself." Simmel, "The Metropolis and Mental Life," 2.

51. "From one angle, life is made infinitely more easy in the sense that stimulations, interests, and the taking up of time and attention present themselves from all sites and carry it in a stream which scarcely requires any individual efforts for its ongoing. But from another angle, life is composed more and more of these impersonal cultural elements and existing goods and values which seek to suppress peculiar personal interests and incomparabilities." Simmel, "The Metropolis and Mental Life," 9.

52. Simmel, "The Metropolis and Mental Life," 2.

Chapter 10

1. Paul Shepheard, *The Cultivated Wilderness: Or, What Is Landscape?* (Cambridge, MA: MIT Press, 1997), 233. "Be aware of the strategy that governs what you do" (233).

2. "Don't call it Romanticism!" may have first been said in the imperative by Charlene Spretnak in her book *The Resurgence of the Real: Body, Nature, and Place in a Hypermodern World* (New York: Routledge, 1997), 131–135.

3. "Old scenarios of environmental meltdown today seem naïve in their apocalyptic dress. People have become too sophisticated—and also too habituated to crisis—to respond simplistically to end-of-the-world alarm sounding. This increase in crisis sophistication does not, of course, mean that the crisis itself has gone away. Instead it indicates that more and more, crisis has become part of the milieu in which people, even in the crisis-denying United States, increasingly dwell." Lawrence Buell, *From Apocalypse to Way of Life: Environmental Crisis in the American Century* (New York: Routledge, 2004), 75–76.

4. See Jared Diamond, *Guns, Germs, and Steel: The Fates of Human Societies* (New York: Norton, 1997). "Westward runs the course of empire" is attributed to the eighteenth-century philosopher George Berkeley, after whom the hometown of a far western university was named.

5. Grey Brechlin, *Imperial San Francisco: Urban Power, Earthly Ruin* (Berkeley: University of California Press, 1999). 13–70.

6. Donald Worster, *The Ends of the Earth: Perspectives on Modern Environmental History*, ed. Donald Worster (New York: Cambridge University Press, 1988), 289.

7. Worster, *The Ends of the Earth,* 293. Reflecting back on the first decades of "doing environmental history," Worster emphasized how artifacts had been excluded from the origins of this field. "The social environment, the scene of humans only interactive with each other in the absence of nature, is therefore excluded. Likewise is the built or artifactual environment, the cluster of things that people have made and which can be so pervasive as to constitute a kind of 'second nature' around them" (292).

8. It would be inaccurate to trace such generalizations about criticism to any single source, but if I had to pick one, it would be Dave Hickey's famous essay "Air Guitar," in *Air Guitar: Essays on Art and Democracy* (Los Angeles: Art Issues Press, 1997), 163–171.

9. Lawrence Buell, *The Future of Environmental Criticism: Environmental Crisis and Literary Imagination* (Oxford: Blackwell, 2005), 17–28. Here, in "The Environmental Turn Anatomized," Buell investigates the social construction of science, in the manner of Latour, but recognizes that literature also identifies fields and creates terms.

10. "Raymond Williams' *The Country and the City* (1973) . . . has been praised as a masterpiece of ecocriticism *avant la lettre* [i.e., ontologically]. . . . Williams more closely anticipated later literature-and-environment studies, in his keen interest in the facts of environmental history, [and] in literature's (mis)representation of them." Buell, *The Future of Environmental Criticism,* 14, 15.

11. William Cronon, *Nature's Metropolis: Chicago and the Great West* (New York: Norton, 1991). Of interest to smart-city technologists, Cronon's work also shows how the network effects involved in those transformations led toward abstraction and acceleration of strategy, such as futures trading.

12. John Stilgoe, *Metropolitan Corridor: Railroads and the American Scene* (New Haven: Yale University Press, 1983), 21–45. In graduate seminars, I still use this classic study of the railroads as "gateway" onto network studies.

13. Adam Markham, *A Brief History of Pollution* (London: Earthscan, 1994), 54.

14. Lewis Mumford, *Technics and Civilization* (New York: Harcourt, Brace, 1936), 167, 255.

15. "The debate around pollution revolves around definitions, ethics, and attitudes toward nature. What is pollution? Is it wrong to pollute? Who is responsible? Who or what is suffering and how?" Markham, *A Brief History of Pollution*, 25.

16. See Chris Leinberger, "The Next Slum," *Atlantic*, March 2008. Leinberger's essay soared to the top of the *Atlantic*'s most-mailed list and stayed there for several weeks.

17. David Owen, "Green Manhattan: Why New York Is the Greenest City in the U.S.," *New Yorker*, October 18, 2004.

18. Sitting down with a print book still has a place among the experiences of reading. A paper book will still work twenty years hence, and anytime you want it. Meanwhile, there is nothing better than books for lining the walls of your house.

19. Geoffrey Nunberg, "Farewell to the Information Age," in *The Future of the Book*, ed. Geoffrey Nunberg (Berkeley: University of California Press, 1996), 103–138. The word *information* began to appear in the nineteenth century.

20. Geoffrey C. Bowker, *Memory Practices in the Sciences* (Cambridge, MA: MIT Press, 2006), 29–31.

21. Alberto Manguel, *A History of Reading* (London: Harper Collins, 1996), 191–192.

22. See Alberto Manguel, *The Library at Night* (New Haven: Yale University Press, 2008).

23. Or, according to "a set of rules that describe our reactions to technologies" by the author of *The Hitchhiker's Guide to the Galaxy*: "1. Anything that is in the world when you're born is normal and ordinary and is just a natural part of the way the world works. 2. Anything that's invented between when you're fifteen and thirty-five is new and exciting and revolutionary and you can probably get a career in it. 3. Anything invented after you're thirty-five is against the natural order of things." Douglas Adams, *The Salmon of Doubt: Hitchhiking the Galaxy One Last Time* (London: Macmillan, 2002), 95.

24. Beth Potier, "Data Overload Nothing New: Too Much Information Has Vexed Scholars for Centuries," interview of Ann Blair, *Harvard Gazette*, February 13, 2003.

25. Gottfried Leibniz, as quoted in Ann Blair, *Too Much to Know: Managing Information before the Modern Age* (New Haven, Yale University Press, 2010).

26. See, for example, Katherine Elison, *Fatal News: Reading and Information Overload in Early Eighteenth Century Literature* (London: Routledge, 2006). "I find that 18th century experience with the proliferation of texts and expansion of its communication systems is unique in the history of media" (2).

27. Elison, *Fatal News*.

28. Jonathan Swift, "Tale of the Tub," as quoted in Nunberg, "Farewell to the Information Age."

29. Andro Linklater, *Measuring America: How an Untamed Wilderness Shaped the United States and Fulfilled the Promise of Democracy* (New York: Walker, 2002), 8–10.

30. By the late seventeenth century, information emerges as a concept, and almost immediately it is imagined as a physically and psychologically threatening entity, and once material and immaterial, capable of overwhelming the human body and intellect.

31. Octave Mirabeau, as quoted in Ferguson, *Paris as Revolution*, 131.

32. See Henkin, *City Reading*, for many vivid examples of nonbook text.

33. Brian Cowan, "Inventing the Coffee House" and "Penny Universities," in *The Social Life of Coffee: The Emergence of the British Coffeehouse* (New Haven: Yale University Press, 2005), 79–112. Word searches through the diary Pepys kept from 1660 to 1669, now fully available online, don't turn up much on signage, streetscapes, handbills, posters, newspapers, or coffeehouses, for that matter, even though the first coffeehouses had appeared in London and Oxford well before then.

34. Henkin, *City Reading*, 72.

35. Henkin, *City Reading*, 70–75.

36. Henkin, *City Reading*, ix.

37. Ferguson, *Paris as Revolution*, 38–39.

38. Carolyn Merchant, *The Columbia Guide to American Environmental History* (New York: Columbia University Press, 2002), xiii.

39. See David Hackett Fischer, *Albion's Seed: Four British Folkways in America* (New York: Oxford University Press, 1989).

40. Linklater, *Measuring America*, 160–175.

41. David Nye, *America as Second Creation: Technology and Narratives of New Beginnings* (Cambridge, MA: MIT Press, 2003), 1.

42. Linklater, *Measuring America*, 13–17, 181, on the impact of Gunter's 66-foot chain as a standard in land surveying.

43. The distinction of broadcast radius persisted for at least half a century. In the days before public radio (NPR was not founded until 1970, by which time radio was often presumed dead), you could tell you were approaching New York by the rare sound of WQXR's classical music on your car radio.

44. The website http://earlyradiohistory.us/ offers many essays from and about the period after World War I when radio became a civilian and commercial technology. They show the origins of such now essential notions as transmitting recorded music and personal wireless voice communication.

45. See Bill Bishop, *The Big Sort: Why the Clustering of Like-Minded America Is Tearing Us Apart* (Boston: Houghton Mifflin, 2008).

46. Klingman, *Brandscapes*, 65.

47. John Brinkerhoff Jackson, *American Space: The Centennial Years, 1865–1876* (New York: Norton, 1972), 30. "To be doing something where it had never been done before and in an altogether new manner—this was worth recording and celebrating" (30).

48. "These [village lawns and centennial groves] survive to remind us of a nationwide movement to create a new kind of communal space, a more natural, more open environment; each was an anonymous testimonial for a love of American landscape." Jackson, *American Space*, 37.

49. "In an age like our own where sensitivity to environment has been exalted to the status of virtue, there are doubtless many features of those attempted reforms which we find uncongenial: a reliance on mechanical solutions, a tendency to subdue nature rather than cooperate with it, an unawareness of interaction instead of opposition between man and the world surrounding him" Jackson, *American Space*, 37.

50. "The causes were several, but perhaps it can be said that for a variety of reasons the slum, instead of being popularly defined by the kind of people who lived there, was redefined by the kind of houses to be found there, by the environment." Jackson, *American Space*, 223.

51. Economist Herman Daly introduced the word *intrinsic* in drawing a fundamental, yet complementary contrast between instrumental and intrinsic forms of value. There exists, he asserted, an awareness that a continuum has value in itself. A forest need not be "land of many uses," but can be "for" nothing, and just exist.

52. Richard Nisbett, *The Geography of Thought: How Asians and Westerners Think Differently. . . And Why* (New York: Free Press, 2003), 89–92, on the "focal fish" experiment.

Chapter 11

1. See Adam Markham, *A Brief History of Pollution* (London: Earthscan, 1994).

2. See www.darkskies.org.

3. Amid the abandoned swaths of Detroit, artists have been putting up billboard images of starlit skies.

4. Peter Blake, *God's Own Junkyard: The Planned Deterioration of America's Landscape* (New York: Holt, Rinehart and Winston, 1964), 69.

5. Thomas Cusack Company v. City of Chicago, 242 U.S. 526 (1917).

6. See John Jakle and Keith Sculle, *Signs in America's Auto Age: Signatures of Landscape and Place* (Iowa City: University of Iowa Press, 2004).

7. William K. Ewald and Daniel R. Mandelker, *Street Graphics: A Concept and a System* (Washington, DC: American Society of Landcape Architects Foundation, 1971), a model municipal sign codes book.

8. KCET television studios, "Billboard Confidential," October 9, 2008, http://www.kcet.org/shows/socal_connected/content/environment/billboards/. LA Weekly called these "billboard/supergraphic/major advertising/ Blade Runner districts." Many cities have focused their efforts on the creation and governance of signage districts, even if in Los Angeles those inevitably bring comparisons with that iconic local tale, *Blade Runner* (1982).

9. Wayne Franklin, foreword to Jakle and Sculle, *Signs in America's Auto Age*, xiv.

10. "Outdoor Advertising Control Language Guide," www.fhwa.dot.gov/realestate/oacguide.htm.

11. "The State of Electric Signage," www.signindustry.com.

12. Shenk, *Data Smog*, 31.

13. Norbert Wiener, *The Human Use of Human Beings: Cybernetics and Society* (Boston: Houghton Mifflin, 1950), 21–34.

14. Cory Doctorow, "How Copyright Broke," in *Content: Selected Essays on Technology, Creativity, Copyright, and the Future of Future* (San Francisco: Tachyon, 2008), 83–88. Available for free at http://craphound.com/content/download/.

15. Tom Hayes, "Next Up for the Internet: The Attention Rights Movement," TomBomb.Com, January 22, 2009, http://tombomb.typepad.com/tombomb/2009/01/next-up-for-the-internet-the-attention-rights-movement.html.

"Consider the following manifesto I am offering for the movement. Hard to disagree with these simple dignities:

- I am the sole owner of my attention.
- I have a right to compensation for my attention, value for value.
- Demands on my attention shall be transparent.
- I have a right to decide what information I want. And don't want.
- I own my click stream and all other representations of my attention.
- My email box is an extension of my person. No one has an intrinsic right to send me mail.
- Attention theft is a crime."

16. Hayes, "The Attention Rights Movement."

17. Stephen Carter, *Civility: Manners, Morals, and the Etiquette of Democracy* (New York: Basic Books, 1998), 4–6. Carter opened with a parable of the shift from the etiquette of transcontinental railroad travel, when "travelers understood their obligation to treat each other well," and when etiquette guides were published and read, to the age of the private automobile, when "we travel both long and short distances surrounded by metal and glass and

the illusion that we are traveling alone. The illusion has seeped into every crevice of our public and private lives, persuading us that sacrifices are no longer necessary" (4).

18. Carter, *Civility,* 11.

19. Carter, *Civility* 191–193.

20. Charles Ess, "Floridi's Philosophy of Information and Information Ethics: Current Perspectives, Future Directions," *Information Society* 25 (2009): 89–96.

21. Luciano Floridi, "Foundations of Information Ethics," in Himma and Tavani, eds., *The Handbook of Information and Computer Ethics,* 12: "Entropy here refers to any kind of destruction, corruption, pollution, and depletion of informational objects (mind, not of information), that is, any form of impoverishment of being."

22. Lewis Hyde, *Common as Air: Revolution, Art, and Ownership* (New York: Farrar, Straus and Giroux, 2010), 44. Alas I started long before finding this, but now anyone who wants to think more about commons could hardly do better than to start here.

23. Hyde, *Common as Air,* 38. See also Charlotte Hess and Elinor Ostrom, "Introduction: Overview of the Knowledge Commons," in Hess and Ostrom, eds., *Understanding Knowledge as a Commons,* 3–26.

24. According to Blaise Pascal, "tyranny is the wish to obtain by one means what can only be had by another." As quoted in Hyde, *Common as Air,* 221.

25. Elinor Ostrom, *Governing the Commons: The Evolution of Institutions for Collective Action* (Cambridge: Cambridge University Press, 1990), 7.

- Clearly defined boundaries should be in place
- Rules in use are well matched to needs and customs
- Individuals affected by these rules can usually participate in modifying these rules
- The right of community members to modify their own rules is respected by external authorities
- A system for self-governing members' behavior has been established
- A graduated system of sanctions is available

• Community members have access to low cost conflict resolution mechanisms

• Nested enterprises, that is, appropriation, provision, monitoring, sanctioning, conflict resolution, other governance activities—are organized in a nested structure with multiple layers of activities.

26. Paul Hawken, *Blessed Unrest: How the Largest Movement in the World Came into Being and Why No One Saw It Coming* (New York: Viking Press, 2007), 1, 2.

27. David Bollier, "Growth of the Commons Paradigm," in Hess and Ostrom, eds., *Understanding Knowledge as a Commons*, 29.

28. Hess and Ostrom, "Introduction: Overview of the Knowledge Commons," 4.

29. For perhaps a distant but clear second to the visibility of Hardin's *Tragedy of the Commons*, see Carol Rose, "Comedy of the Commons: Commerce, Custom and Inherently Public Property" (1986), reprinted as chapter 5 in her *Property and Persuasion: Essays on the History, Theory and Rhetoric of Ownership* (Boulder, CO: Westview Press, 1994), 105–162.

30. Hyde, *Common as Air*, 235.

31. Cory Doctorow, "Andy Warhol Is Turning in His Grave," (Manchester) *Guardian*, November 13, 2007, http://www.guardian.co.uk/technology/2007/nov/13/pop.art.copyright. All twenty-some mentions of "commons" in Doctorow's recent book-scale compilation, *Content*, are of creative commons.

32. Ivan Illich, "Silence Is a Commons." *CoEvolution Quarterly*, Winter 1983, http://www.preservenet.com/theory/Illich/Silence.html.

33. James Boyle, "The Second Enclosure Movement and the Construction of the Public Domain," *Law and Contemporary Problems* 66 (Winter–Spring 2003): 33.

34. Hyde, *Common as Air*, 110.

Chapter 12

1. Nick Bilton, *i live in the future and here's how it works: why your world, work, and brain are being creatively disrupted* (New York: Crown, 2010), 103–132, on "Suggestions and Swarms."

2. See Winifred Gallagher, *Rapt: Attention and the Focused Life* (New York: Penguin Press, 2009), 189–202, for the relation of well being, neuroplasticity, and getting beyond industrial blasé.

3. As of January 2012, "peak distraction" wasn't yet a meme. Indeed, very few uses were findable in a search.

4. David Henkin's expression "city reading" could catch on here, and his method of studying everyday uses of printed information technology at street level might well work on the augmented city as well.

5. I am aware that these two very different halves of the book may seem in need of a keystone, but I think that keystone is something about sensibility and is therefore best left to each reader to imagine. This short "synopsis and retrospect" has declared why the two parts are here together. The hindsight that this chapter anticipates should reveal best how much of all this was true.

Epilogue

1. Ivan Illich, "Silence Is a Commons," *Co-Evolution Quarterly*, Winter 1983, 2, http://www.preservenet.com/theory/Illich/Silence.html.

2. Illich, "Silence Is a Commons," 1.

3. Sara Maitland, *A Book of Silence* (Berkeley: Counterpoint, 2008), 3.

4. Maitland, *A Book of Silence,* 3.

5. Maitland, *A Book of Silence,* 12.

6. Arvo Pärt, as quoted in Mark Prendergast, *The Ambient Century: From Mahler to Trance: The Evolution of Sound in the Electronic Age*, with a preface by Brian Eno (London: Bloomsbury Press, 2001), 95.

7. Mother Teresa, as quoted in Carter, *Civility*, 289.

8. Douglas Hofstadter, *Gödel, Escher, Bach: An Eternal Golden Braid* (New York: Vintage Books, 1979), 156.

9. "Cage had professed that his favorite music was when everything was still, when nothing was attempted. The very sounds of his everyday environment were 'poetry to his ears.'" Prendergast, *The Ambient Century,* 1.

10. See George Michelsen Foy, *Zero Decibels: The Quest for Absolute Silence* (New York: Scribners, 2010).

11. See chapters 3 and 4 for detailed arguments by and quotations of the Kaplans and others on attention restoration and effortless attention.

12. Ken Burns, Dayton Duncan, et al., *The National Parks: America's Best Idea*, PBS, 2008.

13. Garret Keizer, *The Unwanted Noise of Everything We Want: A Book about Noise* (New York: PublicAffairs, 2010), 179–188. Despite being wealthy and not just ordinary Americans themselves, members of the "Wise use movement" for motorized recreation cast their opponents as elitists.

14. Keizer, *The Unwanted Noise of Everything We Want,* 131–164, "Their World Too," in which living with noise has become a widespread social inequity.

15. Keizer, *The Unwanted Noise of Everything We Want,* 65–66.

16. Keizer, *The Unwanted Noise of Everything We Want,* 162.

17. Keizer, *The Unwanted Noise of Everything We Want,* 15.

18. Keizer, *The Unwanted Noise of Everything We Want,* 213. Since I have quoted so extensively from Keizer's wonderful book, let me clarify that I discovered it long after I had written drafts of this essay. The piece by Illich was the initial catalyst. Keizer has done a very good service, and I hope he likes this book too.

Name Index

Subject Index